T0183335

SpringerBriefs in Animal Sciences

More information about this series at http://www.springer.com/series/10153

Peter Vršanský

Cockroaches from Jurassic sediments of the Bakhar Formation in Mongolia

Springer

Peter Vršanský
Earth Science Institute
Slovak Academy of Sciences
Bratislava, Slovakia

Institute of Zoology
Slovak Academy of Sciences
Bratislava, Slovakia

Paleontological Institute
Russian Academy of Sciences
Moscow, Russia

ISSN 2211-7504   ISSN 2211-7512   (electronic)
SpringerBriefs in Animal Sciences
ISBN 978-3-030-59406-0   ISBN 978-3-030-59407-7   (eBook)
https://doi.org/10.1007/978-3-030-59407-7

This Springer imprint is published by the registered company Springer Nature Switzerland AG
The registered company address is: Gewerbestrasse 11, 6330 Cham, Switzerland

Acknowledgements

I thank Prof. Alexandr Pavlovich Rasnitsyn (PIN RAS, Moscow) and Jun Hui Liang (Museum of Natural History, Tianjin) for their valuable reviews and advice. I deeply recognise the collective of the Artropod Laboratory of the Paleontological Institute of the Russian Academy of Sciences for collecting and revealing all specimens for study and for kind acceptance during the study. Especially acknowledged are Irina Dmitrievna Sukatcheva for extensive logistic support, Dmitry Evgenevich Shcherbakov, Dmitry Vladimirovič Vasilenko and Jan Hinkelman (IZ SAS Bratislava, Zagreb) for technical help and Alexandr Georgievich Ponomarenko for organising expeditions. I thank Tatiana Kúdelová (Comenius University, Bratislava) for advising phylogenetic analyses. This work was supported by the Slovak Research and Development Agency under the contracts no. APVV-0436-12; VEGA 2/0042/18 and by UNESCO-Amba/MVTS supporting grant of Presidium of the Slovak Academy of Sciences and its interacademic exchanges.

Abstract

Daohugou in China is the most important Middle Jurassic reference fossil Lagerstätte in the world. To understand its context I studied unselectively collected, roughly coeval 12 different assemblages of 16 genera, including 6 new genera and 32 species from the geographically adjacent Bakhar locality in Mongolia (n= 1,178; 23.4% of all insects). Dominant families were Caloblattinidae, Liberiblattinidae, Blattulidae and Raphidiomimidae depending on assemblage. Mesoblattinidae and Chresmodidae were rare; Fuziidae, diverse in Daohugou were absent. *Ano da* sp. n. (Liberiblattitnidae; n= 83) revealed high forewing venation variability ($CV_{veins\ at\ margin}$= 13.65). Bed 275/1 with 334 identified cockroaches housed in 16 morphologically distinct species (and at least another five indicated by pronota) belongs to the most diverse localities in the 310 Ma order history. Alternative theories explaining the high diversity are rapid changes of the local biota or a prolonged sedimentation bed time. Diversity, if existing locally, is only comparable to the tropical amber of Myanmar, which supplies living associations of extant rainforests and Upper Jurassic sedimentary assemblages of the Karabastau Fm. At the same time, other beds of the same locality are standard in terms of diversity. Indigenous (endemic) genera were absent, which is surprising taking into consideration 19 reported indigenous Jurassic genera. Cockroach fossils include 8 complete specimens, 307 isolated hind wings, 47 isolated pronota and the remainder were mostly fragmentary forewings (4.0-30 mm) occasionally preserved with sophisticated coloration. Composition of respective beds within Bakhar differs significantly and was both higher and lower than that occurring at stable Daohugou with collected complete specimens only. Within-bed (208; 268; 275) taphonomy is consistent except 208/4 with different species and immature individuals. Phylogenetic analysis reveals that cockroaches are in adequate evolutionary stages: primitive or advanced families were absent, 4 (indigenous at current state of knowledge) genera radiated, 3 genera represent FOD, while no LOD suggests that the assemblage had generally rather advanced characters. Deformities represent only about 5% of wings including isolated clavi. Cockroaches, sharing 7 widespread genera and a single advanced one out of a total of 16 genera respectively at Daohugou, but 11 of 16 (11/77; including 7 advanced ones) with Karabastau alone indicate rather Upper than Middle Jurassic age, on stage comparable (not earlier) than Karabastau Fm of Kazakhstan. This age is further supported by different paleogeographical findings in the province of Karabastau.

Contents

Introduction to Bakhar Cockroaches

Cockroaches are ecologically and paleontologically important insects easily recognisable due to their "fat" body along with endosymbionts used in nitrogen fixation. With exceptions of predators and some specialised groups, they are distinct in having hypognathous heads concealed under a pronotum. Their common recognizable preservation as fossils is the forewing (rigid tegmina) representing one of the most common remains among terrestrial records. Wings can be distinguished from other insects on the basis of elevated posterobasal parts called clavus, which protects essential parts of the body posterior to pronotum.

With ~110,000 collected specimens, they are the most abundant terrestrial insects in the sedimentary fossil record, important for the stratigraphy (Schneider 1977, 1978a–c, 1980a, b, 1982, 1983, 1993), phylogeny (Vršanský 2002, 2010) and ecology (Vršanský et al. 2002, 2013, 2016, 2017, 2018a, b, 2019a–d) of ancient and understanding of living biotas (Vidlička 2001).

The Mesozoic is the most studied interval of their evolution, with roughly 50,000 catalogized samples, of which about 50% have already been examined. Most important sites within the Triassic period were formalised by Fujiyama (1973), Papier et al. (1994), Papier and Grauvogel-Stamm (1995), Vishniakova (1998), Fang et al. (2013).

The Mesozoic is also important with significant reference points represented by K1 and K2 Lebanese and Myanmar ambers (Vršanský 2003, 2004; Ross and Grimaldi 2004; Poinar and Buckley 2006; Anisyutkin and Gorochov 2008; Poinar 2009a, b; Vršanský and Bechly 2015; Sendi and Azar 2017; Poinar and Brown 2017; Šmídová and Lei 2017; Vršanský and Wang 2017; Li and Huang 2018a, b; Qiu et al. 2019a, b; Kočárek 2018a, b; Bai et al. 2016, 2018; Gao et al. 2019; Podstrelená and Sendi 2018; Mlynský et al. 2019; Šmídova 2020, Hinkelman 2020, Hinkelman and Vršanská 2020, Sendi et al. 2020a, b).

The Jurassic has vast collections, including ~20,000 specimens that remain formally unevaluated (Handlirsch 1906; Vršanský and Ansorge 2007; Vršanský 2008), although nearly 3,000 Karabastau cockroaches have already been surveyed

(Vishniakova 1971, 1973; Vršanský 2009; Liang et al. 2018, 2019; Vršanský et al. 2019). Jurassic cockroaches were formalised by Germar (1839), Heer (1852, 1964, 1965), Giebel (1856), Oppenheim (1888), Brauer et al. (1889), Haughton (1924), Hong (1980, 1983, 1997), Lin (1982, 1985, 1986), Handlirsch (1939), Bode (1953), Fujiyama (1974), Whalley (1985), Zhang (1986), Wang (1987), Martynova (1951), Ren et al. (1995), Hong and Xiao (1997) and Martin (2010).

In Mongolia and adjacent areas, sedimentary records of J/K and $K1$ were revised (Vishniakova 1964, 1968, 1980, 1982, 1993; Vršanský 2004, 2008).

Regarding the middle Jurassic, only about 200 Daohugou specimens (of roughly 8,000 collected from the site; 5,000 in the CNU collection—Gao et al. 2019) were described within formally erected taxa (Liang et al. 2009, 2012a, b; 2018, 2019; Wei et al. 2012, 2013; Guo and Ren 2011; Gao et al. 2019; Vršanský et al. 2009, 2012). Middle Jurassic Kubekovo ($n = 23$) in Russia was also partially evaluated (Vishniakova 1982, 1985). Bakhar is another Lagerstatte, preliminary dated as terminal Middle Jurassic or basal Late Jurassic.

The Early Jurassic record contains primitive Jurassic forms (revised by Vršanský and Ansorge 2007; Martin 2010) while the Late Jurassic is already a time period with all major Cretaceous cockroach lineages already diversified (Vršanský 2007; 2008a, b, Vishniakova 1968, 1982, 1983, 1985, 1971, 1973; Vršanský et al. 2019), and the calibration dating of advanced (Cretaceous) cockroach with ootheca diversification makes Bakhar to be a milestone. Later, advanced cockroach assemblages were formed in Mongolian Shar-Teg, a Cretaceous-type, although these are now considered Barremian due to presenting the diversification point with occurrences of wing deformities associated with first representatives of lineages such as mantodeans (Vršanský 2004, 2005; Vršanský et al. 2017). Earlier ootheca-holders are considered on the basis of a preserved forewing from the basalmost Jurassic of Suffield in the USA (Huber et al. 2003).

These fossils represent the same insect order (Vršanský et al. 2009), with cosmetical differences from living representatives. These include mostly externally protruding ovipositor of females, absence of egg case and presence of central ocellus. In contrast to living (and Palaeozoic) cockroaches, size variability was minimal in Mesozic cockroach species. No living families existed during the herein studied interval. Differences from Palaeozoic representatives were most likely also minimal, but the evidence is missing for the nitrogen fixation.

Development of extinct and living cockroaches is similar, they are "incomplete" (it is surely complete for those reaching adulthood), and all living cockroaches lay eggs in rigid egg-case, ootheca, protecting them from predators and parasites (although specialised parasites Evaniidae exist). Extinct cockroaches laying a single egg have serious consequences making their offspring without protection.

This study further exemplifies the appearance of main cockroach lineages (Blattulidae and Caloblattinidae), which were extremely conservative during their ca. 200 Ma existence and did not produce any descending taxon. At the same time, rare Liberiblattinidae and Mesoblattinidae appeared, and these families later gave birth to numerous offshoots including taxa with brand new morpho- and ecotypes such as termites, praying mantises or viviparous taxa (and all modern cockroaches).

Aim of the present study was to systematically describe more than a thousand cockroach fossils (Figs. 1, 2, 3, 4, 5, 6, 7, 8, 9, 10, 11, 12, 13, 14 and 15) from the middle Jurassic locality Bakhar in Mongolia, compare it with the other described and undescribed taxa from Daohugou and Kubekovo and evaluate the ecological, stratigraphical and phylogenetic context (Figs. 16, 17, 18, 19, 20 and 21). It is surprising for a discovered extremely high diversity.

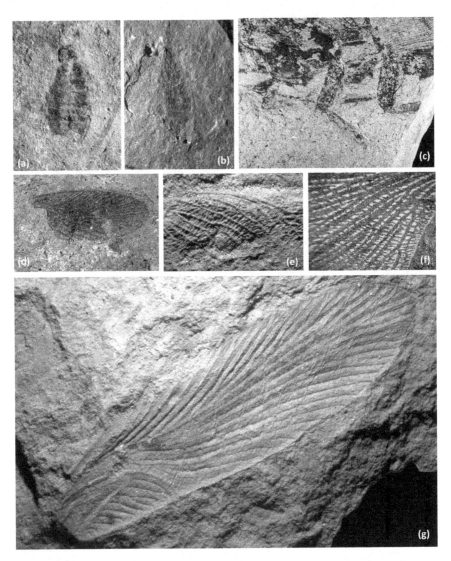

Fig. 1 Examples of cockroach taphonomy along the Jurassic Bakhar profile (PIN 3791/). **a** uniden-tified blattulid larva PIN 3791/106, **b** a complete liberiblattinid specimen PIN 3791/1147 = PIN 3791/1148 (*Hra nie* gen. et sp.n. holotype; body with pronotum and head 10 mm long), **c** articulated body structures including legs in a complete caloblattinid PIN 3791/739 (complete individual ca. 30 mm long), **d** fragmentary blattulid forewing with preserved coloration PIN 3791/1203 (*Blattula bacharensis* sp.n. holotype Bed 238; 5.3 mm), **e** forewing clavus with preserved 3D vein profiles PIN 3791/446 (*Solemnia togokhudukhensis* sp.n. Bed 275/1; 16-mm-long fragment), **f** forewing with fine coloration pattern PIN 3791/960 (*Rhipidoblattina konserva* sp.n.; 7 mm fragment), **g** complete forewing of a predatory Raphidiomimidae PIN 3791/70 (*Raphidiomima chimnata* sp.n. holotype; 17.7/5.0 mm)

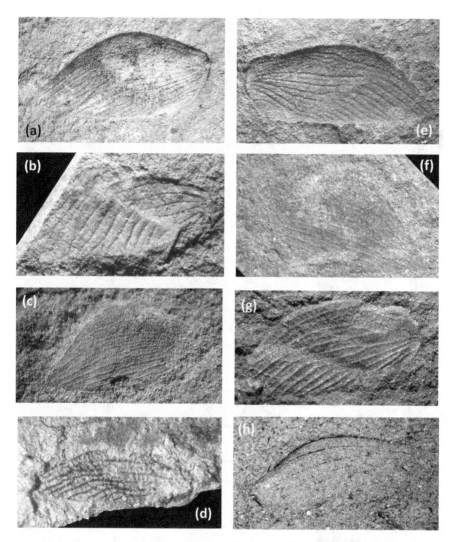

Fig. 2 Isolated Jurassic cockroach forewing clavi of Bakhar locality (PIN 3791/). **a** PIN 3791/424 (*Caloblattina vremeni* sp.n. Bed 275/1; 9 mm long), **b** PIN 3791/65 (*Raphidiomima chimnata* sp.n. Bed 208/2; 4.3 mm long), **c, f** PIN 3791/724 = 731 (*Perlucipecta cosmopolitana* sp.n. holotype Bed 275/1; length 5 mm), **d** PIN 3791/776 (*Raphidiomima krajka* sp.n. holotype Bed 275/1; 3.8 mm long), **e** PIN 3791/600 (*Solemnia togokhudukhensis* sp.n. Bed 275/1; 7.5 mm long), **g** PIN 3791/140 (*Rhipidoblattina sisnerahkab* Bed 268/8; 6.5 mm long), **h** PIN 3791/786 (*Blattula velika* sp.n. Bed 275/1; 5 mm long)

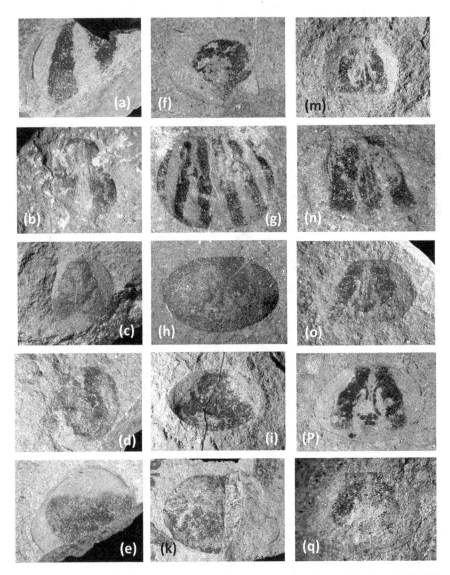

Fig. 3 Isolated Jurassic cockroach pronota of Bakhar locality (PIN 3791/). **a** PIN 3791/453 (length/width 5.0/6.0 mm), **b** PIN 3791/427 (5/4 mm), **c** PIN 3791/1188 (5.0/5.0 mm), **d** PIN 3791/1182 (5.4/5.0 mm), **e** PIN 3791/560 (/5.8 mm), **f** PIN 3791/855 (2.4/4.0 mm), **g** PIN 3791/958 (4.4/5.8 mm), **h** PIN 3791/433 (4/7 mm), **i** PIN 3791/624 (5.5/9.0 mm), **k** PIN 3791/895 (6.0/9.0 mm), **m** PIN 3791/778 (2.9/3.5 mm), **n** PIN 3791/514 (3.8/5.2 mm), **o** PIN 3791/880 (3.1/5.0 mm), **p** PIN 3791/550 (3.9/6.7 mm), **q** PIN 3791/533 (3.7/5.3 mm)

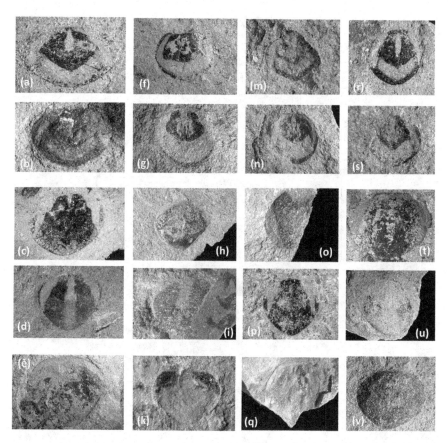

Fig. 4 Isolated Jurassic cockroach pronota of Bakhar locality (PIN 3791/). **a** PIN 3791/534 (2.0/3.2 mm), **b** PIN 3791/1199 (2.3/3.1 mm), **c** PIN 3791/438 ± (1.9/2.3 mm), **d** PIN 3791/669 (5.0/5.5 mm), **e** PIN 3791/432 (5.5/6.0 mm), **f** PIN 3791/805 (3.3/4.0 mm), **g** PIN 3791/1192 (3.5/3.5 mm), **h** PIN 3791/701 (/2.7 mm), **i** PIN 3791/662 (4.0/4.3 mm), **k** PIN 3791/599 (2.3/2.7 mm; *Ano da*), **m** PIN 3791/735 = 714 (2.0/2.1 mm), **n** PIN 3791/812 (4/3.8 mm), **o** PIN 3791/729 (ca. 5.0/6.0 mm), **p** PIN 3791/436 (2.1/2.1 mm), **q** PIN 3791/69 (4 mm; *Raphidiomima chimnata*), **r** PIN 3791/891 (2.6/2.8 mm), **s** PIN 3791/714 = 735 (2.0/2.1 mm), **t** PIN 3791/491 (5.2/7.1 mm), **u** PIN 3791/544 (5.6/6.0 mm), **v** PIN 3791/933 (3.4/4.0 mm)

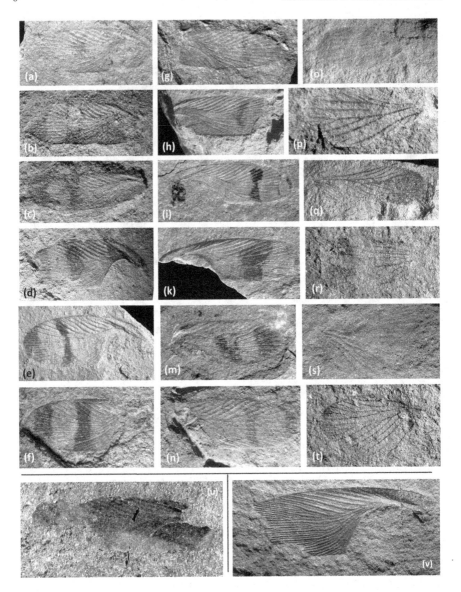

Fig. 5 Isolated Jurassic liberiblattinid cockroach forewings of Bakhar locality (PIN 3791/; Bed 275/1 (most); 275/2 (g); 208/3 (v); 268/4 (s)). **a–o, r** *Ano da* sp.n., **p** *Polliciblattula vana* sp.n., **q, t** *Blattula vulgara* sp.n., **s** *Hra nie* sp.n., **u** *Ano net* sp.n., **v** *Ano nym* sp.n. **a** PIN 3791/877 5.8 mm, **b** PIN 3791/474 (holotype), **c** PIN 3791/569 8/2.7 mm, **d** PIN 3791/415 8.2/4.5 mm, **e** PIN 3791/618 6/2.4 mm, **f** PIN 3791/909 6 mm long, **g** PIN 3791/1189 5.5/1.9 mm, **h** PIN 3791/784 (7.5/2.8 mm), **i** PIN 3791/706 (8 mm long), **k** PIN 3791/804 9/2.5 mm, **m** PIN 3791/663 (6.2 mm long), **n** PIN 3791/895 (6 mm long), **o** PIN 3791/5054 (5 mm long), **p** PIN 3791/606 (holotype; 4.5 mm long), **q** PIN 3791/570 5.7 mm, **r** PIN 3791/617 (4.5 mm long), **s** PIN 3791/129, **t** PIN 3791/692, **u** PIN 3791/842 (holotype; 10.5 mm long), **v** PIN 3791/94 (holotype; 11-mm-long fragment)

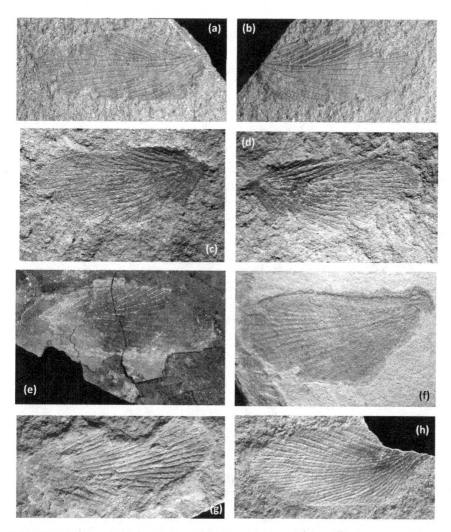

Fig. 6 Isolated Jurassic cockroach forewings of Bakhar locality (PIN 3791/). **a, b** PIN 3791/444 (*Hra disko* sp.n. holotype Bed 275/1; ca. 14/5.5 mm), **c, d** PIN 3791/749 (*Hra disko* sp.n. Bed 275/1; 10.5-mm-long fragment), **e** PIN 3791/1026 (*Praeblattella jurassica* sp.n. holotype Bed 328; 13 mm long), **f** PIN 3791/268 (?*Hra nie* sp.n. Bed 268/14; 17 mm), **g** PIN 3791/63 (*Raphidiomima chimnata* sp.n. Bed 208/2), **h** PIN 3791/907 (*Hra disko* sp.n. Bed 275/1; 6-mm-long fragment)

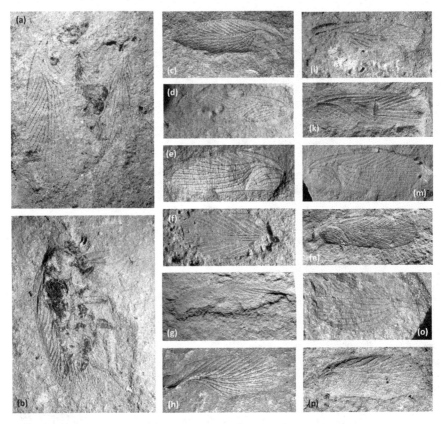

Fig. 7 Bakhar locality cockroaches (PIN 3791/) **a, b** PIN 3791/637, PIN 3791/449 (*Blattula vulgara* sp.n. Bed 275/1), **c** PIN 3791/419 (*Blattula vulgara* sp.n. holotype Bed 275/1; 8.5/2.5 mm), **d** PIN 3791/88 (*Blattula universala* sp.n. Bed 208/2; 9/2.2 mm), **e** PIN 3791/64 (*Blattula flamma* sp.n. holotype Bed 208/2; 6/2 mm), **f** PIN 3791/497 (*Blattula velika* sp.n. Bed 275/1;/3.6 mm), **g** PIN 3791/52 (*Truhla vekov* sp.n. holotype 208/2; 8.7 mm long), **h** PIN 3791/945 (*Hra bavi* sp.n. holotype Bed 275/1; 12/3.6 mm), **i** PIN 3791/102 (*Truhla vekov* sp.n. Bed 208/3; 8.7 mm long), **k** PIN 3791/73 (*Blattula flamma* sp.n. Bed 208/2; 1.5-mm-long clavus), **m** PIN 3791/5051 (*Blattula mikro* sp.n. holotype 275/1; 4.8/1.8 mm), **n** PIN 3791/516 (*Blattula mini* sp.n. holotype Bed 275/1; 4.5/1.3 mm), **o** PIN 3791/123 (*Hra nie* sp.n. Bed 268/4; 5–6 mm), **p** PIN 3791/92 (*Polliciblattula analis* sp.n. holotype Bed 208/2; 4/1.4 mm)

Fig. 8 Articulated Bakhar cockroaches (PIN 3791/). **a** PIN 3791/446 (*Solemnia togokhudukhensis* sp.n. Bed 275/1; 16-mm-long fragment), **b** PIN 3791/270 = 357 (*Rhipidoblattina sisnerahkab* sp.n. Bed 268/14; 13-mm-long left forewing), **c** PIN 3791/959 (*Rhipidoblattina konserva* sp.n. Bed 268/14; 24.5 mm), **d** PIN 3791/111 (*Rhipidoblattina sisnerahkab* sp.n. Bed 268/4; 16 mm), **e** PIN 3791/645 (*Blattula anuniversala* sp.n. Bed 275/1; 12.7 mm), **f** PIN 3791/897 (*Perlucipecta cosmopolitana* sp.n. Bed 275/1; 13-mm-long fragment)

Fig. 9 Bakhar locality cockroaches (PIN 3791/) **a, b** PIN 3791/588, PIN 3791/468 (*Nuur-cala cela* sp.n. Bed 275/1; 28.3 mm), **c, d** PIN 3791/926 (*Nuurcala cela* sp.n. holotype Bed 275/1; 17/6 mm), **e** PIN 3791/105 (*Nuurcala?cela* sp.n. Bed 268/4; 6-mm-long fragment), **f** PIN 3791/127 (*Rhipidoblattina konserva* sp.n. Bed 268/8; 12.5 mm fragment), **g** PIN 3791/456 (*Solemnia togokhudukhensis* sp.n. holotype Bed 275/1; 18/4.7 mm), **h** PIN 3791/721 (*Solemnia togokhudukhensis* sp.n. Bed 275/1; 7-mm-long fragment), **i** PIN 3791/294 (Bed 268/14; 12/4.5 mm), **k** PIN 3791/117 (*Rhipidoblattina* sp.; 14.5 mm)

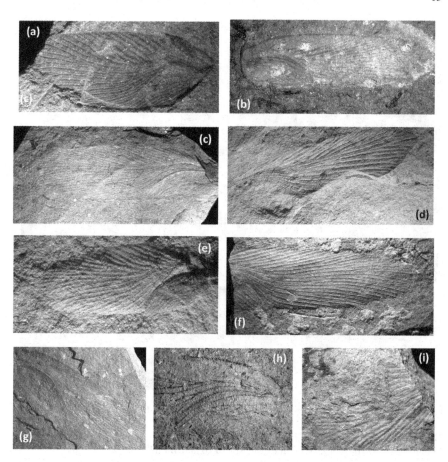

Fig. 10 Bakhar locality cockroaches (PIN 3791/) **a** PIN 3791/582 (*Okras sarko* sp.n. holotype Bed 275/1; 10/3.8 mm), **b** PIN 3791/205 = 290 (*Rhipidoblattina sisnerahkab* sp.n. Bed 268/14; 14.8/3.9 mm), **c** PIN 3791/278 (*Rhipidoblattina konserva* sp.n. holotype Bed 268/14; 22/6.5 mm), **d** PIN 3791/53 (*Raphidiomima chimnata* sp.n. Bed 208/2; 16/4 mm), **e** PIN 3791/574 (*Rhipidoblattina bakharensis* sp.n. holotype Bed 275/1; 14 mm), **f** PIN 3791/496 (*Caloblattina vremeni* sp.n. holotype Bed 275/1; 18/5.2 mm), **g** PIN 3791/976 (*Rhipidoblattina konserva* sp.n. Bed 268/14 24 mm), **h** PIN 3791/508 (*Caloblattina vremeni* sp.n. Bed 275/1; fragment 10 mm long), **i** PIN 3791/63 (*Raphidiomima chimnata* sp.n. Bed 208/2)

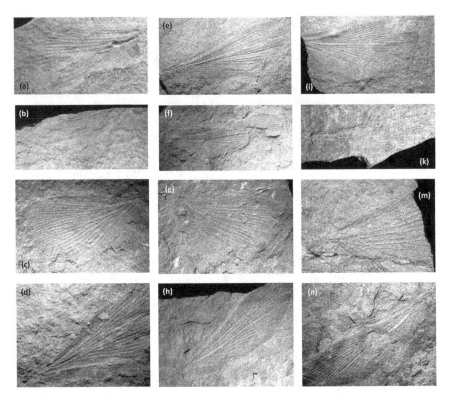

Fig. 11 Complete Bakhar cockroach hindwings (PIN 3791/). **a** PIN 3791/101 10 mm, **b** PIN 3791/445 18 mm, **c** PIN 3791/302 11-mm-long fragment, **d** PIN 3791/607 21 mm, **e** PIN 3791/722 22 mm, **f** PIN 3791/136 20 mm, **g** PIN 3791/988 13 mm, **h** PIN 3791/315 20 mm, **i** PIN 3791/593 15 mm, **k** PIN 3791/121 14.5 mm, **m** PIN 3791/310 10 mm, **n** PIN 3791/285 17 mm

Fig. 12 a Complete specimen of common Middle Jurassic cockroach *Ano da* sp.n. (Bakhar Bed 275/1) revealing associated pronotum uncommon in profile (PIN 3791/599; 7.1/2.2 mm), **b** *Blattula aberrans* Vishniakova, 1982 PIN 1214/1 from Kubekovo, Siberia

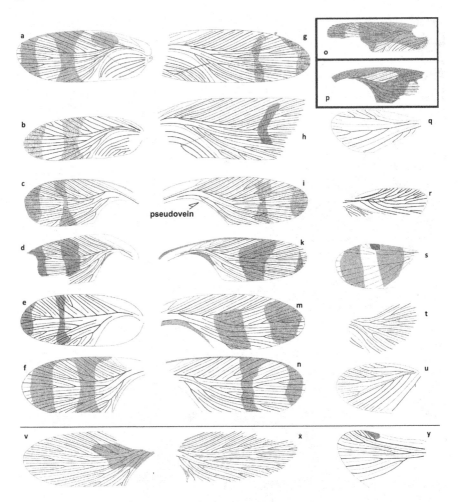

Fig. 13 Illustrations of Bakhar Liberiblattinidae and Blattulidae (PIN 3791/; all except **g** (275/2); **p** (208/3); **t** (268/8) from Assemblage 275/1). **a** *Ano da* sp.n. PIN 3791/877 5.8 mm, **b** *Ano da* sp.n. holotype PIN 3791/474 7.5 mm, **c** *Ano da* sp.n. PIN 3791/618 6 mm, **d** *Ano da* sp.n. PIN 3791/415 8.2 mm, **e** PIN 3791/294, **f** *Ano da* sp.n. PIN 3791/909 6 mm, **g** *Ano da* sp.n. PIN 3791/1189 5.5 mm, **h** *Ano da* sp.n. PIN 3791/784 7.5 mm, **i** *Ano da* sp.n. PIN 3791/706 8 mm, **k** *Ano da* sp.n. PIN 3791/804 9 mm, **m** *Ano da* sp.n. PIN 3791/663 6.2 mm, **n** *Ano da* sp.n. PIN 3791/895 6 mm, **o** *Ano net* sp.n. PIN 3791/842 10.5 mm, **p** PIN 3791/94 *Ano nym* sp.n. holotype 11 mm, **q** *Polliciblattula vana* sp.n. holotype PIN 3791/606 4.5 mm, **r** *Blattula vulgara* sp.n. PIN 3791/570 7.8 mm, **s** PIN 3791/617 *Ano da* sp.n. 4.5 mm, **t** PIN 3791/129 *Hra nie* sp.n. fragment, **u** PIN 3791/692; **v** PIN 3791/907 *Hra disko* sp.n. 1.6 mm, **x** PIN 3791/444 *Hra disko* sp.n. holotype 14 mm, **y** PIN 3791/5054 *Ano da* sp.n. 5 mm

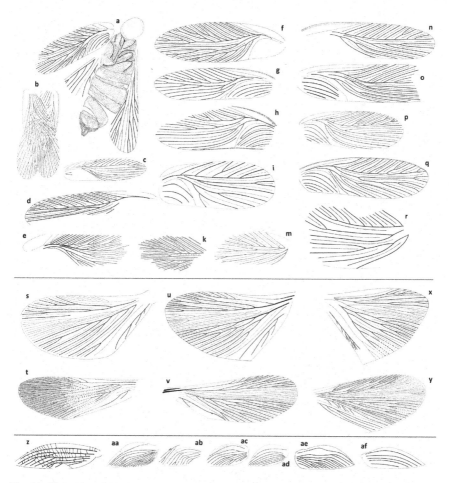

Fig. 14 Illustrations of Bakhar cockroaches (PIN 3791/). **a** *Hra nie* sp.n. holotype PIN 3791/1147 = 1148 (268/14) body with pronotum and head 10 mm, **b** *Blattula anuniversala* sp.n. holotype PIN 3791/645 (275/1) 12.7 mm, **c** *Dostavba pre* sp.n. holotype PIN 3791/405 8 mm (268/19), **d** *Truhla vekov* sp.n. holotype PIN 3791/52 (208/2) 8.7 mm, **e** *Hra bavi* sp.n. holotype PIN 3791/945 (Bed 275/1) 12 mm, **f** *Blattula vulgara* sp.n. holotype PIN 3791/419 (275/1) 8.5 mm, **g** *Blattula universala* sp.n. holotype PIN 3791/88 (Bed 208/2) 9 mm, **h** *Blattula flamma* sp.n. holotype PIN 3791/64 (208/2) 6 mm, **i** *Polliciblattula analis* sp.n. holotype PIN 3791/92 (208/2) 4 mm, **k** *Blattula velika* sp.n. PIN 3791/497 (275/1) 3.6 mm, **m** *Blattula flamma* sp.n. PIN 3791/109 (208/2) 6.5 mm estimated, **n** *Truhla vekov* sp.n. PIN 3791/102 (208/3) 8.7 mm, **o** *Blattula flamma* sp.n. PIN 3791/73 (208/2) 1.5 mm-long-clavus, **p** *Blattula mikro* sp.n. holotype PIN 3791/5051 (275/1) 4.8 mm, **q** *Blattula mini* sp.n. holotype PIN 3791/516 (275/1) 4.5 mm, **r** *Hra nie* sp.n. PIN 3791/123 (268/4) 5–6 mm, **s** PIN 3791/302 11 mm, **t** *Rhipidoblattina konserva* sp.n. PIN 3791/285 17 mm, **u** ?*Hra nie* sp.n. PIN 3791/268 (268/14) 17 mm, **v** *Rhipidoblattina konserva* sp.n. PIN 3791/315 (268/14) 20 mm, **x** PIN 3791/988 (268/14) 13 mm, **y** *Rhipidoblattina bakharensis* sp.n. PIN 3791/593 (275/1) 15 mm, **z** *Raphidiomima krajka* sp.n. holotype PIN 3791/776 (275/1) 3.8 mm, **aa** *Caloblattina vremeni* sp.n. PIN 3791/424 (275/1) 9 mm, **ab** *Raphidiomima chimnata* sp.n. PIN 3791/65 (208/2) 4.3 mm, **ac** *Rhipidoblattina sisnerahkab* sp.n. PIN 3791/140 (268/8) 6.5 mm, **ad** *Perlucipecta cosmopolitana* sp.n. holotype PIN 3791/724 = 731 (275/1) length 5 mm, **ae** *Solemnia togokhudukhensis* sp.n. PIN 3791/600 (275/1) 7.5 mm, **af** *Blattula velika* sp.n. PIN 3791/786 (275/1) 5 mm

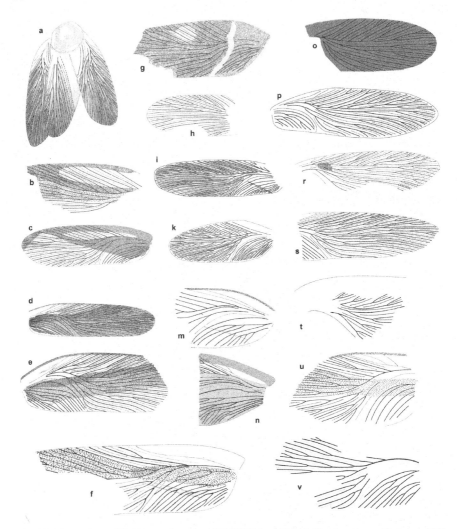

Fig. 15 Illustrations of Bakhar cockroaches (PIN 3791/). **a** *Rhipidoblattina sisnerahkab* sp.n. PIN 3791/270 = 357 holotype (268/14) 13-mm-long left forewing, **b** *Nuurcala cela* sp.n. PIN 3791/588 (275/1) 28.3 mm, **c** *Rhipidoblattina* sp. PIN 3791/117; 14.5 mm, **d** *Rhipidoblattina sisnerahkab* sp.n. PIN 3791/205 (268/14) 14.8 mm, **e** *Nuurcala cela* sp.n. holotype PIN 3791/926 (Bed 275/1) 17 mm, **f** *Solemnia togokhudukhensis* sp.n. holotype PIN 3791/456 (275/1) 18 mm, **g** *Okras sarko* sp.n. holotype PIN 3791/582 (Bed 275/1) 10 mm, **h** *Raphidiomima chimnata* sp.n. PIN 3791/60 (208/2), **i** *Rhipidoblattina konserva* sp.n. holotype PIN 3791/278 (268/14) 22 mm, **k** *Rhipidoblattina bakharensis* sp.n. holotype PIN 3791/574 (275/1) 14 mm, **m** *Caloblattina vremeni* sp.n. PIN 3791/508 (275/1) fragment 10 mm long, **n** *Solemnia togokhudukhensis* sp.n. PIN 3791/721 (275/1) 7-mm-long fragment, **o** *Blattula bacharensis* sp.n. holotype PIN 3791/1203 (238) 5.3 mm, **p** *Raphidiomima chimnata* sp.n. holotype PIN 3791/70; 17.7 mm, **r** *Raphidiomima chimnata* sp.n. PIN 3791/53 (208/2) 16 mm, **s** *Caloblattina vremeni* sp.n. holotype PIN 3791/496 (275/1) 18 mm, **t** *Nuurcala ?cela* sp.n. PIN 3791/105 (268/4) 6-mm-long fragment, **u** PIN 3791/294 (268/14) 12 mm, **v** PIN 3791/127 *Rhipidoblattina konserva* sp.n. (268/8) 12.5-mm-long fragment

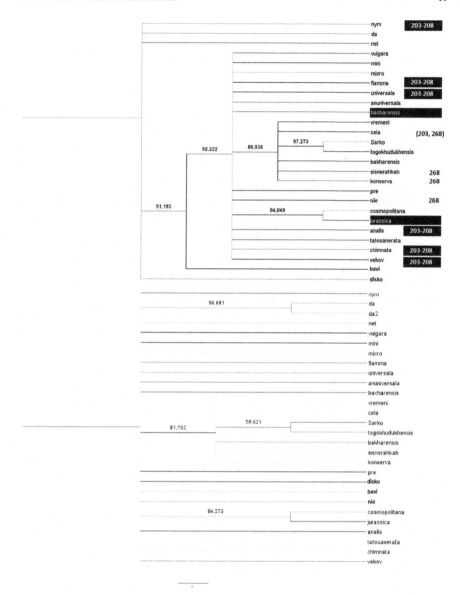

Fig. 16 Phylogenetic analysis was performed with most parsimonious trees of the cockroach species from all the Bakhar profile (species lacking complete forewing information are disregarded here) heuristic search with 100 and 1,000, ACCTRAN as well as TBR algorithms. Characters were treated as unordered, unweighted, numbers reveal bootstrap values

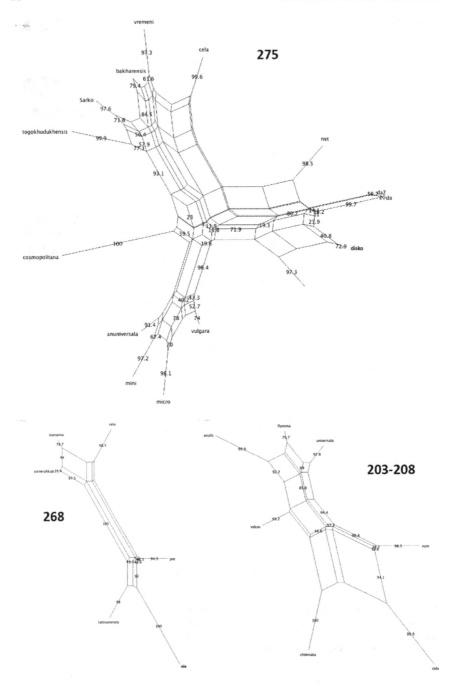

Fig. 17 Bayesian network analyses of the cockroach species from the major Bakhar assemblage (203 + 208; 268; 275) profile (species lacking complete forewing information and species from 328 are disregarded here), neighbour-net algorithm. Numbers along edges represent bootstrap values (only supports above 50 displayed)

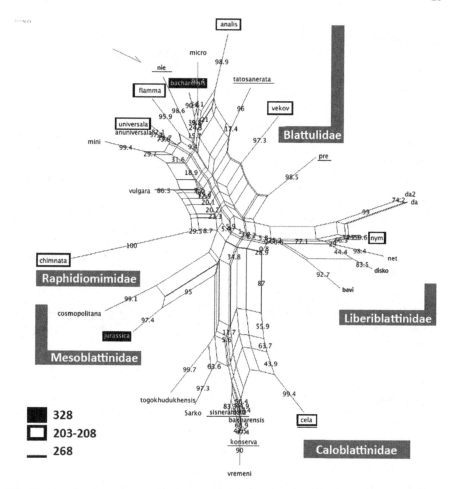

Fig. 18 Bayesian network analysis of the cockroach species from all the Bakhar profile (species lacking complete forewing information, neighbour-net algorithm. Numbers along edges represent bootstrap values (only supports above 50 displayed))

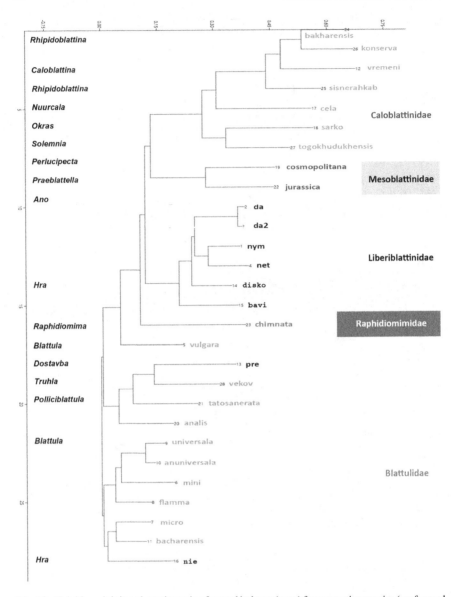

Fig. 19 Neighbour joining clustering using Jaccard indexes (axes) for respective species (performed in PAST)

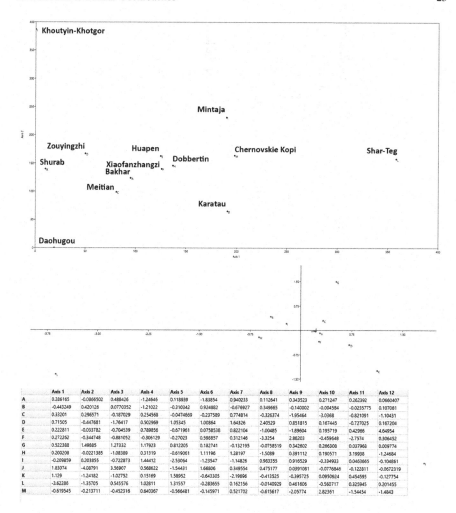

	Axis 1	Axis 2	Axis 3	Axis 4	Axis 5	Axis 6	Axis 7	Axis 8	Axis 9	Axis 10	Axis 11	Axis 12
A	0.286165	-0.0866502	0.488426	-1.24646	0.118939	-1.83854	0.940233	0.112641	0.343523	0.271247	0.262392	0.0660407
B	-0.443249	0.420126	0.0770352	-1.21022	-0.310342	0.924882	-0.676927	0.349665	-0.140002	-0.004584	-0.0235775	0.107081
C	0.33201	0.296571	-0.187029	0.254568	-0.0474669	-0.237589	0.774814	-0.326374	-1.95464	-3.0368	-0.821091	-1.10431
D	0.71505	-0.447681	-1.76417	0.502969	1.05345	1.00864	1.64326	2.40529	0.851815	0.167445	-0.727025	0.167204
E	0.222811	-0.033782	-0.704539	0.789856	-0.671961	0.0758538	0.822104	-1.00485	-1.89604	0.195719	0.42966	4.64954
F	0.272262	-0.344748	-0.881052	-0.306129	-0.27023	0.598857	0.312146	-3.3254	2.86203	-0.459648	-2.7574	0.306452
G	0.522388	1.49885	1.27332	1.17923	0.812205	0.182741	-0.132195	-0.0758519	0.342602	0.266308	0.037968	0.009774
H	0.200208	-0.0221385	-1.08389	0.31319	-0.619061	1.11196	1.28197	-1.5089	0.391112	0.190571	3.19938	-1.24684
I	-0.209859	0.203855	-0.722873	1.44412	-2.53064	-1.23547	-1.14826	0.983355	0.916529	-0.334933	0.0463665	-0.104861
J	1.83074	-4.08791	3.56907	0.568622	-1.54431	1.66806	0.349554	0.475177	0.0391081	-0.0776846	-0.122811	-0.0672319
K	1.129	-1.24182	-1.02752	0.15169	1.58952	-0.643305	-2.19696	-0.413525	-0.395725	0.0950924	0.454595	-0.127754
L	-3.62286	-1.35705	0.545576	1.02811	1.31557	-0.283655	0.162156	-0.0140929	0.461606	-0.560717	0.325945	0.201455
M	-0.619545	-0.213711	-0.452316	0.640367	-0.566481	-0.145971	0.521702	-0.615617	-2.05774	2.82361	-1.54454	-1.4843

Fig. 20 Correspondence ordination coordination analyses, (top) without *Falcatussiblatta* and (bottom) with *Falcatussiblatta* (with indexes provided in the table). Numbers represent the same assemblages in both analyses (performed in PAST)

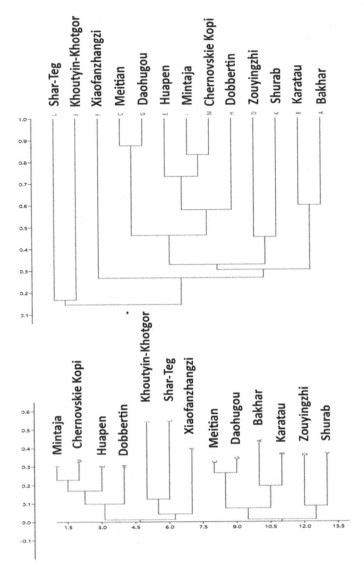

Fig. 21 Hierarchical clustering (top) and Neighbour joining method (bottom) with Jaccard indexes (performed in PAST)

Methods of Study Fossil Cockroaches

Material ($n = 1{,}178$ catalogised cockroaches (1,182 in the table due to divided samples) among 5,141 insects; 19 additional specimens are under the same numbers as two or more impressions) was collected by the Arthropod Laboratory of the Paleontological Institute of the Russian Academy of Sciences (formerly USSR) in 1974 (V.N Yakovlev, the head of the expedition); and in 1979 and 1980 (A.G. Ponomarenko, the head of expeditions); in the Bajan–Chongor Aimag, 12 km NE of Cace–Ula in Mongolia, and catalogised and housed under numbers PIN 3791/n.

Specimens 3791/420, 436, 443±, 458, 460, 463, 464, 472, 500, 515, 542, 557, 565, 587, 639, 648, 653, 664, 675, 682, 700, 708, 715p, 754, 767, 781, 782, 787, 794, 795, 796, 802, 819, 829, 831, 833, 838, 839, 846, 853, 860, 889, 894p, 928, 947, 948, 950 (275/1); 992, 242, 248, 257, 231, 303, 1025, 5192, 975±p, 1028, 1032, 1040, 1066p, 1088, 1114, 1138, 1139, 1167, 1141 (268/14); 388 (268/19); 96 (208/3) were determined as Blattaria and deposited in the cockroach collection, but probably do not belong to this order or had been damaged during the collection/deposition (so that I was unable to locate cockroaches on these samples unequivocally).

Unidentifiable Blattaria (Assemblage 275/1): 3791/416, 417±, 429±, 430±FFH, 435±, 437, 442, 457, 466, 467, 475, 494, 503, 521, 526, 527, 536, 543, 549, 564, 577, 580, 581, 605fb±, 613, 626, 627, 636, 638ff, 640±, 649, 657c, 674, 681, 683, 687, 688, 696c, 699, 707c, 713, 716, 727, 743, 770c, 771, 783, 788, 790, 799, 823, 849bf, 866, 867ff, 882, 903, 920c, 922, 924, 930, 941, 943, 944c, 957 (forewings); 3791/504, 507, 551, 553, 611±, 614±, 620, 625, 656, 666, 668, 725, 736, 744, 755, 760, 768, 769, 774, 775, 780, 818, 845, 861, 878, 876, 856, 912, 918, 955 (hindwings); 3791/568, 598leg, 612±p, 667b, 694p (bodies); 3791/432, 450, 454, 459, 477, 510, 529, 530, 538, 546, 548, 583, 597±, 613, 621, 635, 654, 761, 785, 789, 792, 796, 827, 830, 832, 836, 848, 857, 858, 865, 873, 874, 875, 886, 892, 898, 902, 906, 915, 921, 927, 935, 951, 956 (unidentifiable). **SUM: F 65 (C6; FFH1; BF2, FF2), H 30, B5(P3, LEG1); SUM + 44 = 144**.

(Assemblage 275/2): 3791/1185, 1177±, 1183, 1184, 1187, 1190, 1194, 1195, 1196p, 1197. **SUM forewings: 9; bodies 1(P)**. Pronota: 1192 = 1200, 1182, 1199.

P. Vršanský, *Cockroaches from Jurassic sediments of the Bakhar Formation in Mongolia*, SpringerBriefs in Animal Sciences, https://doi.org/10.1007/978-3-030-59407-7_2

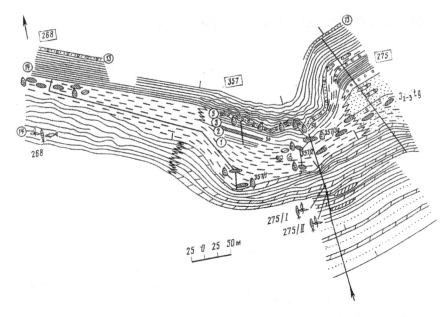

Fig. 22 Stratigraphy of basal horizon 275 and central-upper 357; example of facial replacement of eastern psephytes with western aleurolites and pellites in 275, 257 and 268. After Sinitsa 1993

Isolated undetermined pronota (length/width): 3791/426 ± (2.2/3.2 mm), 432 (5.5/6.0 mm), 433 (4/7 mm), 436 (2.1/2.1 mm), 438 ± (1.9/2.3 mm), 453 (5.0/6.0 mm), 491 (5.2/7.1 mm), 498, 514 (3.8/5.2 mm), 520 (4.1/6.9 mm), 533 (3.7/5.3 mm), 534 (2.0/3.2 mm), 544 (5.6/6.0 mm), 550 (3.9/6.7 mm), 560 (/5.8 mm), 624 (5.5/9.0 mm), 662 (4.0/4.3 mm), 669 (5.0/5.5 mm), 684 (2.0/2.0 mm), 701 (/2.7 mm), 714 = 735 (2.0/2.1 mm), 729 (ca. 5.0/6.0), 762 (4.5/6 mm), 773 (5.0/6.4 mm), 778 (2.9/3.5 mm), 800 (/ ca. 4 mm), 805 (3.3/4.0 mm), 824 (3.3/5.0 mm), 855 (2.4/4.0 mm), 880 (3.1/5.0 mm), 885 (/2.1 mm), 891 (2.6/2.8 mm), 895 (6.0/9.0 mm), 911 (5.2/5.9 mm), 912 (4.0/3.8 mm), 933 (3.4/4.0 mm), 936, 958 (4.4/5.8 mm). **SUM: 38 (275/1)**.

Stratigraphically, as a stratotype, it belongs to Bakhar Member (Devyatkin et al. 1975), with Bakhar complex containing Togokhuduk (assemblages 268, 275) and Ortsogsk (assemblages 208) rock strata and Khoutiin-Khotgor complex with Bajan-Ul rock statum (assemblages 327, 328) (Sinitsa 1993) (Fig. 22).

Photographs were made using a Leica binocular lens with a Nikon pix. Camera is manually attached to the right ocular. Some of the photographs were manually combined.

Drawings represent redrawn photographs checked under the microscope for venation details. Some of them were adjusted for coloration using ADOBE Photoshop. Wing nomenclature follows the Comstock-Needham system (1898–1899).

Abbreviations used: SC, subcosta, R, radius, RS, radius sector, M, media, Cu, cubitus (CuA, anterior, CuP, posterior), A, anal veins, CW, cross veins, IC, intercalary veins, f, forewing, ff, both forewings, h, hindwing, hh, both hindwings, p, pronotum, c, isolated clavus.

The study of the specimens was performed mainly during 2003–2005 in the Arthropoda Laboratory, PIN RAS. Material was evaluated blindly in respect of the geological setting.

Specimens were categorised within respective species on the basis of size and character match disregarding their relation to the respective beds. Consequently, composition within respective beds was compared and variability considered. This way, some taxa without unique variability, but individually identified as distinct, were merged with species occurring in the same beds.

Original reports list molluscs, ostracodes *Darwinula sarytirmensis* Sharapova, 1947 *D.* cf. *impudica* Sharapova, 1947 (Sinitsa 1985), fishes and part of a pterosaur (Bachurina and Unwin, 1995). Insects are the most common representatives of Bakhar oryctocenoses (21 orders, ca. 50 families and ca. 1,000 species): most common are beetles ($n = 1,885$), followed by cockroaches ($n = 1,165$), Hemiptera ($n = 862$), Mecoptera and Orthoptera, unexpectedly common were caddis cases, since aquatic orders are rare ($n \sim 100$) (Ponomarenko 2019).

For phylogenetic analysis (Figs. 16–18), I computed the most parsimonious trees (see discussion paragraph parsimony analysis) in PAUP * 4.0b8 (Swofford 2003) using a heuristic search, ten random additional taxon replicate, the accelerated transformation optimisation algorithm (ACCTRAN), as well as the three bisection–reconnection branch-swapping (TBR) algorithm. Characters were treated as unordered and unweighted. A 50% majority rule consensus tree was constructed from most parsimonious trees found during the heuristic search. Branching reliability was assessed by the bootstrap method with 10, 100 and 1000 replicates. A phylogenetic network is constructed in SplitsTree 4 (neighbour-net algorithm—Bryant and Moulton 2004) with bootstrap analysis (1,000 replicates) in effect.

Ordination analysis (Figs. 19–21) was performed using PAST.

Geological Settings and Environments

Environments of the Bakhar localities (in detail elaborated by Ponomarenko 2019; translated and shortened herein for overview of conditions): basal fluvial conglomerates consist of transported boulders and pebbles, and fluvial sandstones and aleurolites without organic debris. Rhythmic profile represents sediments of rather large lake remnants with diverse sedimentary sources. Lake transgression was rapid, with sediments becoming more fine-grained and the depth was sufficient for preservation of aquatic and terrestrial insects (Assemblage 275). Aquatic insects were dominantly represented by emphemerans (Leptophlebiidae Banks, 1900; *Mesoneta* Brauer et al. 1889), rarely by aquatic beetles, bugs and chresmodids. Terrestrial insects are most common herein. Sediments of the central part formed as the condition of the delta changed at adjacent parts of the lake. The characteristic of sediments is mass fragments of thick-walled bivalve shells, conchostracans and green algae oogonia. Terrestrial organisms are represented by plant debris. In fine-grained 268/3, 4, 8, there are aquatic bugs, backswimmers (Notonectidae), corixid *Bakharia gibbera* and aquatic beetles, without plecopteran larvae (imago present). Terrestrial insects are more rare than in 275. The upper part of the profile is characterised by dark grey and black fine-grained argillites (Assemblages 268/14, 15) with abundant dispersed coalised organics, leaves, seeds and terrestrial insects. Aquatic organisms are represented only by paleoniscid fishes (and terrestrial adult stages of odonates). This zone represents deposits of small dislocated parts of the lake, densely covered by forest at the banks; dispersed organics were transported from land as a product of debris decay (Sinitsa 1993).

In Ortsag time, the area of sediment forming widens to West and South. Basally, weakly rounded gravel material predominates, eroded by temporary flows in a rapid manner, but without too much caving in from the hill margin. In sandstones, most stable crystal predominates; thus, the banks and adjacent heights were not high, but penplenised with progressive erosion. In gravel, shale and an effusive occur and Palaeozoic limestones were missing, forming a contemporary Western margin of

Bakhar sediments, and thus, the main flow was from the East and South. Sedimentary fans were preserved from Western and Eastern sides of the structure and to the South, a lake with aleurolite sedimentation was formed. The lake was inhabited with rare ostracods and diverse, abundant insects: *Bakharia gibbera, Shuragobia altaica* (in lower horizons), *Tersus*, liaditids and hydrophilids, *Baga bakharica*, chironomids, chaoborids, fish (only scales preserved). Abundant terrestrial insects were also preserved. The lake was continuously transgressing so the sediment becomes more fine-grained and carbonised as terrigenous input decreased. By the end of Ortsag time, the lake remained only in the South without filling up with sand. Pelit filling was absent, and a wide sapropele bed formed a 10 m thick with semirocky coal. From inhabitants of algae sources of sapropele were caddis cases from sand dust. Southeast from this lake, isolated or semiisolated small lakes existed where transport of clastics was abundant, with allochtonous coalised layers. Forests at banks were abundant, plants commonly fossilised, but insects are rare. Aquatic insects are exclusively represented by *Terrindusia* and *Folindusia* (Sukatsheva 1994).

There was a significant time difference between deposition during Ortsag and Bayangul times and Ortsag sediments were dislocated and partially eroded. Consequently, shallow and extensive Bayangul lake originated, with cycles of algal limestone and dark grey bituminous finely laminated argillites (papershells). Water filled with *Khoutynia, Haenbaea badamgaravae*, chaoborids and caddisflies. Insects differed from Dzhargalant and Khamar-Khoburin formations, without a single shared species. Ortsogon insects form a single association with Early and Middle (Miomoptera, *Mesoneta, Mesotaeniopteryx, Cyclothemis*, Shurabellidae, *Mesocixiella, Dysmorphoptila*, Volopinae and Protogryllinae), and also Late Jurassic taxa. The later are more common (Mesopentacoridae, Karataidae, Ephialtitidae, *Xyelula*, numerous artematopoids, Eucinetidae, *Heteromera*, Chrysomelidae-Protoscelinae, advanced Empididae and Rhagionidae, Raphidioptera, *Falsispeculum* and others. FOD are caddis cases, but larvae of Early and Middle Jurassic ephemerans are rare (limited basalmost layers), larvae of plecopterans are absent, and abundant are late Jurassic corixids. Thus, the Bakhar series appears younger than the Dzhargalant and Khamarkhobur formations.

Tectonic activity occurs near the epicentrum with vast earthquakes, revealing smashed blocks exposing different layers, making it impossible to trace lateral facial changes.

Seasonal terrestrial forest at banks of the lake and in adjacent swampy lowlands (studied in detail by Kostina et al. 2015), so-called Tsagan-Ovoo flora contains 32 megafossil plant taxa belonging to horsetails, ferns, cycadaleans, ginkgoaleans (including indigenous *Ginkgo badamgaravii* Kostina et Herman 2015; *Pseudotorellia gobiense* Kostina et Herman 2015, and *Pseudotorellia mongolica* Kostina et Herman 2015), leptostrobaleans, conifers and gymnosperms of unknown systematic affinity. This flora is typical of the West Siberian Province of the Siberian Region, the North Chinese Province is more southwards. It characterises warm temperate humid conditions (Kostina and Herman 2016).

Nevertheless, if kalligrammatids also formed associations with gymnosperms, the widespread presence of *Kalligramma* and *Kalligrammula* in the Middle Jurassic– Lower Cretaceous of Europe and Asia could be due to the similarity of floras in the low and mid-latitudes of Laurasia. All representatives of both genera are known from the territory of the Euro-Sinian paleofloristic province, which is characterised by the abundance of Bennettitales (significantly with "flowers") and Cheirolepidiaceae, and scarcity of Czekanowskiales and Ginkgoaceae (Vakhrameev 1991).

New Assemblages of the Bakhar Locality

Systematic Paleontology

Blattaria Latreille, 1810
 Corydioidea Saussure, 1864
 Liberiblattinidae Vršanský, 2002
 Type genus: *Liberiblattina* Vršanský, 2002. Karabastau Formation.
 Composition: *Brachyblatta, Elisamoides, Entropia, Eublattula, Gurvanoblatta, Kazachiblattina, Kurablattina, Leptolythica, Liberiblattina, Spongistoma Stavba*.
 Stratigraphic range: Early Jurassic—Late Cretaceous.
 Geographical range: cosmopolitan.
 Diagnosis (modified after Vršanský 2002). Forewing with regular venation with terminal dichotomisation limited to the clavus. SC field narrow with SC long and branched. R field narrow with R ending prior to wing apex. M and Cu sigmoidal, M reaching wing apex. CuP strongly curved. Anal veins branched mostly in apical third. Hindwing with fan-like pleating with possible reduction of number of pleating veins, R1 and RS differentiated, pterostigma might occur; M richly branched; CuA with five or more branches, usually secondarily dichotomised. Female with short external ovipositor.

Ano **Gen.N**.

 Type species: *A. da* sp.n. described below.
 Composition: Besides the type species, *A. net* sp.n. described below.
 Stratigraphic range: Middle Jurassic.
 Geographic range: Laurasia (indigenous to Bakhar).
 Differential diagnosis: *Ano* is a highly variable taxon, differing from other representatives of the family in distinct rather simple coloration forming a simple single-dot pattern (other representatives of the family are either without coloration like *Stavba*, with simple macula like *Elisamoides* or with more sophisticated patterns

P. Vršanský, *Cockroaches from Jurassic sediments of the Bakhar Formation in Mongolia*, SpringerBriefs in Animal Sciences, https://doi.org/10.1007/978-3-030-59407-7_4

like *Liberiblattina*). Differences occur in the unparallel margins of rather short and wide forewings with distinct pseudovein (see labelled Fig. 13i; characteristic for mantodeans). Venation is usually characterised with a basally branched rather short SC, but sometimes terminal branchelets are present or SC can be rarely even simple; R slightly sigmoidal with R rarely secondarily branched, RS might be differentiated; M curved, not sigmoidal; CuA with basally differentiated two main stems, A branched, sometimes terminally; CuP fluent. For detailed descriptions, see species *A. da* combined with straight fore margin and the more extensive coloration of *A. net*.

Systematic remarks: The genus is closely related to the type genus *Liberiblattina* Vršanský, 2002, including high congruence in forewing coloration but *Liberiblattina* has additional pale stripe along clavus and additional small pale area on anterior margin. The differences in venation were likely caused by the size (monotypic *Liberiblattina ihringovae* is slightly larger 11 mm and more).

Elisamoides Vršanský, 2004, is similar in posessing two-branched CuA and has wide wings with parallel margins. A few specimens (such as *K. mintajaensis* Martin, 2010, specimen WAM 08.95) of *Kurablattina* Martin, 2010, are very similar (possessing, different to most specimens, two-branched CuA) but with a short clavus.

Gurvanoblatta Vishniakova, 1986, with a short CuA, straight anterior, ascending posterior margins and maculated coloration, is very different. *Brachymesoblatta* Vršanský, 2002 is also structured (miniaturised) entirely differently (tertiary branched R, different coloration).

Eublattula differs in having a simple SC and wider veins. *Kazachiblatttina* Vršanský, 2002 (*K. asiatica* Vishniakova, 1968) has elongated forewings with non-reduced venation and lacks coloration.

The new genus is excluded to represent the family Umenocoleidae in spite of the close ressemblence with the most primitive taxon *Vitisma* Vršanský, 1998, due to lack of synapomorphies (sclerotised forewing) and presence of synapomorphies with other groups (mantodeans), such as the presence of pseudovein. Nevertheless, a close relative of *Ano* from Myanmar amber (I. Koubová and T. Mlynský, in preparation) does not possess raptorial forelegs and it is thus not directly related to praying mantises.

This type of coloration forming an "eye" or a "dot" is missing in middle Jurassic of Daohugou or in earlier localities, and instead is characteristic of undescribed upper Jurassic cockroaches from Karabastau and for Cretaceous *Vitisma* and *Cretaholocompsa* Martínez-Delclós, 1993.

Derivation of name: *ano* is Slavic for Yes. Gender neutrum.

Ano da **sp.n**. (Figs. 4k; 5a–o, r; 12a; 13a–n, s, y)

Holotype: 3791/474. A complete forewing. Assemblage 275/1.
Paratypes: 3791/415±, 423±, 442, 471, 473, 476, 488, 489, 493, 499, 501, 502, 513, 571, 518, 531, 538, 540, 545, 554±, 559, 562, 569, 572, 575. 595±, 599hpf ± sc2, 618±, 642, 661, 663, 689c, 695, 698, 703, 706, 710, 711, 718, 719, 730, 733, 737, 743, 747±, 750ffh, 752, 757, 772, 784, 793, 797, 801, 804, 806, 810, 816ffhh,

822, 870, 871, 877, 863, 868, 884, 888, 893c, 905fh, 917, 919, 929, 938, 940, 943 = 953, 909, 5054hf, 5055 (forewings); 3791/447, 552, 592, 608±, 617±, 764, 879 (hindwings). All assemblage 275/1.

Additional material: 3791/1173±, 1189, 1198 (forewings, assemblage 275/2).

Differential diagnosis: It differs from colored species of Liberiblattinidae in the coloration being dark only sporadically on the forewing membrane, restricted to two horizontal stripes forming pale "dot" among them, a basal macula and a colored anterior margin; forewing comparatively short and wide.

Description: Very small species (forewing length under 10 mm and 1.9–4.5 mm wide). Pronotum (Fig. 4k) nearly triangular, ca 2.4 mm wide and 2 mm long with dark coloration restricted to margins near the widest side. Forewing margins mostly unparallel; poor dark coloration forming white fenestra. SC might be very short (but specimens with long SC occur), R is curved and often very short (sometimes straight and reaching apex). The so-called RS (it is unclear whether RS in cockroaches corresponds to original RS) differentiate only in some specimens. M sigmoidally curved, often reaching apex, sometimes reduced to a single vein. Cu branched and differentiated into two stems. Clavus wide, sharply curved, A might branch or simple. Hind wing very simple, with simple SC; R1 (4–5 with first vein stemming basally) and RS (3–4) differentiated, with small pterostigma; coloration might include striation, while specimen 599 (both hindwings attached to a forewing and pronotum) seems to lack any coloration. M consistently two simply dichotomed branches (four veins meeting margin). CuA with six veins meeting margin, stem tertiary branched in one of the wings; CuP simple. A1 with blind basal stem and terminal branchings.

Systematic remarks: Formally, it is impossible to differentiate the three specimens 3791/1173±, 1189, 1198 originating from assemblage 275/2 into a separate species although these specimens are apparently smaller (5.5 mm compared with 5.8–10.5 mm; mean 7.8 mm) and possibly represent a separate sibling species. The species (or genus) is extremely variable (as most Liberiblattinidae).

Variability analysis based on Table 1 places the specimen 3791/1189 (275/2) within the set of other specimens, but the number of radial branches is lower (8) than any other measured specimens from 275/1 (9–16, ave = 11.7); while set of R + M (14 vs. 13–22, ave = 17.2) and R + Cu (15 vs. 14–22, ave = 17.5) are compensated. Thus, the specimen 1189 cannot be statistically discriminated from others, although there are some differences, here considered for laying at the population level.

Derivation of name: *da* is Russian for Yes. *Anoda* reffers to ἄνοδος (anodos): an electrode through which conventional current flows into a polarised electrical device. It is output releasing from area negative particles of electrical charge. Gender neutrum.

Character of preservation: forewings 80 (c2, fhp1, fh2, ffhh1, ffh1), hindwing 7. 3791/895 is missing in list, as it is listed among isolated pronota possibly due to sample division.

Table 1 (1) means assemblage 275/1; (2) 275/2. Forewing venation variability of *Ano da*

Spec	l (mm)	w (in mm)	SC (veins)	R	M	CuA	C up	A	R + M	R + Cu	M + Cu	SUM
618(1)	6	2.4	2	10	6	6	1		16	16	12	**25**
663(1)	6.2	2.1	2	12	1	5	1		13	17	6	**21**
706(1)	8	2.6	2	16	5	6	1		21	22	11	**30**
895(1)			2	10	5	5	1		15	15	10	**23**
877(1)	5.8	2.1	3	9	4	5	1	6	13	14	9	**22**
784(1)	7.5	2.8	2	10	7	9	1		17	19	16	**29**
703(1)	8	2.6	2	16	5	5	1		21	21	10	**29**
804(1)	9	2.5	4	12	5	5	1		17	17	10	**27**
569(1)	8	2.7	2	12	5	5	1		17	17	10	**25**
423(1)	10	2.9	2	15	7	6	1		22	21	13	**31**
575(1)	7.9	2.5										
415(1)	8.2	4.5	2	11	5	6	1		16	17	11	**25**
476(1)	6.9	2.2		12	6	5	1		18	17	11	**26**
474(1)	7.5	2.4	3	10	8	6	1	8	18	16	14	**36**
493(1)	8	2.4		10	7	7	1		17	17	14	**27**
554(1)	8.5	2.2		11	6	6	1		17	17	12	**26**
599 (1)	7.1	2.2	3	9	4	7	1	6	13	16	11	**24**
577)1	7.4	2.4	2	10	11	3	1	6	21	13	14	**27**
1189 (2)	5.5	1.9	3	8	6	6	1	6	14	15	12	**24**
n	17	17	14	17	17	17	17	4			17	**17**
AVE	7.67	2.56	2.36	11.47	5.7	5.71	1	6.5	17.18	17.18	11.41	**26.65**

(continued)

Table 1 (continued)

Spec	l (mm)	w (in mm)	SC (veins)	R	M	CuA	C up	A	R + M	R + Cu	M + Cu	SUM
Min	5.8	2.1	2	9	1	3	1	6	13	13	6	17
Max	10	4.5	4	16	11	9	1	8	22	22	16	21
Dev	1.097801	0.552335	0.633324	2.239354	2.084607	1.263166	0	1	2.833622	2.404041	2.346775	3.639045
CV	14.31	21.58	26.84	19.52	36.57	22.12	0	15.38	16.49	13.99	20.57	13.65

Ano net sp.n. (Figs. 5u; 13o).

Holotype: 3791/842. More or less complete forewing. Assemblage 275/1.
Additional material: 3791/881 (forewing fragment). Assemblage 275/1.
Differential diagnosis: New species is on the upper size boundary of the species
A. da. Coloration was more extensive and forewing fore margin was nearly straight
compared to stably curved in *A. da.*
Description: Forewing (10.5 mm) with an exception of subterminal pale macula
and basal posterior area dark. Forewing fore margin straight. SC short, dichotomised.
R (12 simply dichotomised branches meeting margin) slightly sigmoidally curved,
not reaching apex, the so-called RS indistinct. M with long, straight apically branched
and curved branches (7). CuA with two main branches. Clavus long, A (A3)
dichotomised, with ca. 8 veins meeting margin.
Systematic remarks: The species was closely related sibling species of *A. da*
(both CuA with two main stems, curved R with indistinct RS) and differences are
restricted to coloration extent.
Derivation of name: *net* is Russian for no. Gender neutrum.
Character of preservation: two forewings.

Ano nym sp.n. (Figs. 5v; 13p).

Holotype: 3791/94±. Forewing fragment. Assemblage 208/3.
Differential diagnosis: It differs from congeners in extensively colored forewing
with pale coloration reduced to only two maculas, while in other species, in genus
forewing is mostly pale; SC richly branched, with five veins meeting margin and
being slightly larger; forewing ca 14 mm long.
Description: SC very long, with five simply dichotomised veins. R slightly
sigmoidally branched. M with at least four straight long branches. CuA differentiated
into two stems, with at least seven branches meeting margin. CuP long.
Systematic remarks: Very closely related species, mostly differing in extent of
coloration and size. Different is probably also the geochronological stage due to
different bed. Only slight sigmoidally curved R resemble *A. net* rather than *A. da.*
Derivation of name: The English suffix-*onym* (o can be omited) is from
the Ancient Greek suffix-ώνυμον (*ōnymon*), neuter of the suffix ώνυμος (*ōnymos*),
having a specified kind of name, from the Greek ὄνομα (*ónoma*), Aeolic
Greek ὄνυμα (*ónyma*), "name". The form-*ōnymos* is that taken by *ónoma* when
it is the end component of a bahuvrihi compound, but in English, its use is extended
to tatpurusa compounds. Also alluding to Nym server, a pseudonym server which
provides untraceable email addresses. Gender neutrum.
Character of preservation: 1 forewing fragment.

Hra Gen.N.

Type species: *Hra disko* sp.n. described below.

Composition: *Hra bavi* sp.n. described below. Assemblage 275/1. *Hra nie* sp.n. described below. Assemblage 268/14.

Type horizon: Assemblage 275/1.

Stratigraphic and geographic distribution: indigenous to Bakhar.

Differential diagnosis: It differs from all Liberiblattinidae except *Elisamoides* in all main branches (R, M, CuA) strong, sigmoidally curved (R slightly, almost reaching apex); strong intercalaries joined with cross-veins, not fully straight. *Elisamoides* differs in having narrow radial field at base, and simplified sharply descending M; original state of CuA (expanded); and having hindwing pterostigma.

Description: Wing main venation and intercalaries wide, CW present. Forewing base wide, costal area short and wide; SC branched; R field short, RS indistinct due to reduction; M not reduced, standard, covering apex; CuA reduced to few branches within narrow cubital field; clavus normally long (1/3–1/4 of the wing length), A simple. Hindwing long, SC long and simple, R1 and RS differentiated, M straight, reduced to some three veins at margin, CuA secondarily branched. Female external ovipositor short.

Systematic remarks: Significantly larger species (up to 13 mm; compared to less than 10 mm in *E. cantabillingensis* Martin, 2010 (Mintaja) and *E. mantiformis* Vršanský, 2004 (Shar-Teg) of the cosmopolitan J1-K2 common genus *Elisamoides* Vršanský, 2004). Larger size resulted in slight straightening of distinct venation curvature compared with *Elisamoides*.

The general morphology of *Elisamoides* is rather conservative and the present genus is rather similar to *Elisamoides cantabillingensis*, which lacks coloration (to the holotype; specimen WAM 08.116 belongs to another genus, *Falcatusiblatta* Liang, Ren et Shih, 2018—see Liang et al. 2018) than to the type species, which is colored. While branching of M is most probably plesiomorphic compared to other genera, simplification of CuA is unusual (other species retained expanded CuA) and likely corresponds with the reduction due to size.

The species can be categorised within Liberiblattinidae on the basis of close relationship with genus *Elisamoides,* extremely wide veins, all of which were signif-icantly sigmoidally curved, many of them densely dichotomised, and characteris-tically colored forewing basal part. The known species is also characterised by extreme variations (also when regarded that the family Liberiblattinidae contains the most variable representatives) suggesting this genus was extremely plastic and this plasticity was retained during the significant time period. Otherwise, the differ-ences among species are weak. Nevertheless, the significant time period differences suggest the taxa erection was safe.

Type genus *Liberiblattina* differs in being extensively colored and in having long SC in narrower forewing. *Kurablattina* Martin, 2010, is very similar, but has more cross-veins, branched A and narrow R base.

Gurvanoblatta Vishniakova, 1986, has also short CuA, straight anterior and ascending posterior margins and maculated coloration, which is different from *Hra*. *Brachymesoblatta* Vršanský, 2002, is also structured (miniaturised) entirely differ-ently (tertiary branched R, different coloration). *Eublattula* differs in having simple SC and narrow R.

Kazachiblattina Vršanský, 2002, (*K. asiatica* Vishniakova, 1968) has more elongate wing with less curved veins.

Derivation of name: *Hra* is almost a stochastical combination of letters meaning a game, play (Slovak). Gender neutrum.

Hra disko sp.n. (Figs. 6a–d, h; 13v, x).

Holotype: 3791/444±. A forewing fragment. Assemblage 275/1.
Paratype: 3791/749, 907. A forewing fragment. Assemblage 275/1.
Type horizon: assemblage 275/1.
Description: Forewing short (ca. 13 mm long and up to 5 mm wide), with margins nearly parallel. Intercalaries present joined with numerous cross-veins. R with secondarily branched veins, with at least nine veins meeting margin. M curved and expanded, covering apex from both sides; with about seven veins meeting margin. CuA with two stems, together also with about seven veins. Coloration restricted to dark shade in the basal part.

Derivation of name: sometimes more, sometimes less stochastistal combination of letters meaning disco (Slovak). Gender neutrum.

Character of preservation: three complete forewings.

Hra bavi sp.n. (Figs. 7h; 14e).

Holotype: 3791/945 = 939. A forewing fragment.
Paratypes: 3791/753, 815, 844 (forewings). All assemblage 275/1.
Type horizon: assemblage 275/1.
Differential diagnosis: The present species differs from congeners in size, which is conserved at about 10 mm, i.e. much bigger than in *Hra nie*. *Hra disko* is bigger in size, but has no coloration which can differ from new species.

Description: Forewing elongated, ca 12 mm long and 3.6 mm wide. Main veins and intercalaries distinct and very wide. Costal area very wide and short, SC branched. Radial branch wide, strongly curved, radial field wide at the base, R veins secondarily branched, M and CuA expanded, CuP simple, clavus long.

Systematic remarks: The species can be categorised within genus on the basis of standard (wide compared to *Elisamoides*) R base, branched M. This taxon unlikely represents males of *Hra disko*, because coloration dimorphism is unknown in Mesozoic cockroaches. Size difference reaching 100% (from *Hra nie*) also seems too large to expain the population differences. Thus, the erection of a new species appears safe.

Derivation of name: *bavi* is a stochastical combination of letters, meaning to have a pleasure (Slovak). Gender neutrum.

Character of preservation: forewings: 4.

Hra nie sp.n. (Figs. 1b; 5s; 6f; 7o; 13t; 14a, r, u).

Holotype: 3791/1148 = 1147. Complete specimen, positive and negative (female).

Paratypes: 3791/308 (forewings); 3791/1113, 1142 (hindwings). All assemblage 268/14.

Additional material: 3791/123 A forewing fragment. Assemblage 268/4; 3791/129 A hindwing fragment 268/8.

Type horizon: assemblage 268/14.

Differential diagnosis: Significantly smaller species (about 5 mm; compared to less than 10 mm in two already known taxa and over 10 in above described taxon). Smaller size resulted in reduction and simplification of A and also in reduction of the width of veins, which are slightly thin.

Description: Head completely concealed under small transverse pronotum. Forewing length/width 4.5/1.8 mm. Costal area wide, SC very short and widely branched into three veins meeting margin. R strongly sigmoidally curved, not reaching apex, with 13 veins meeting margin, three veins are branched, due to reductions and simplification RS impossible to delimit. M expanded, covers the apex, secondarily branched, with at least seven veins at margin. CuA reduced to three posteriorly curved veins, CuP simple, fluently curved posteriorly. Clavus reaches about a third of the wing, four simple anal veins present. Hindwing slightly longer than forewing, with narrow costal area bearing anteriorly slightly curved simple SC. R1 (3) and RS (5) differentiated. Media simplified to three veins at margin; CuA retained the secondary branchings, totally with seven or more veins meeting margin, CuP simple. A1 present in remigium, with two blind branches, one basal (usual in the group) and one posted centrally (autapomorphy). Body wide with distinct externally protruding ovipositor composed of two valves.

Systematical remarks: The present specimen can be categorised within the genus based on general habitus with very wide veins and intercalaries, and extremely vaulted but wide and short stem of R.

These assemblage (268/14) contain very large hindwings (3791/268, 302, 988±) apparently representing another species, which is not formalised herein due to absence of forewings.

Derivation of name: *nie* is nearly a stochastical combination of letters meaning no (Slovak). Gender neutrum.

Character of preservation: 1 complete specimen with disarticulated legs and broken right forewing; 2 forewings; 3 hidwings.

Caloblattinoidea Vršanský et Ansorge in Vršanský (2000)
Caloblattinidae Vršanský et Ansorge in Vršanský (2000)

Type species: *Blattina mathildae* Geinitz, 1883; lower Toarcian, Dobbertin, Germany.

Composition: *Aktassoblatta* Vishniakova, 1971; *Asioblatta* Vishniakova, 1968; *Etapia* Vishniakova, 1983; *Euryblattula* Martynov, 1937; *Fusiblatta* Hong, 1980; *Ijablatta* Vishniakova, 1983; *Itchetuja* Vishniakova, 1983; *Kemerowia* Vishniakova, 1983; *Samaroblatta* Tillyard, 1919; *Samaroblattula* Martynov, 1937; *Sogdoblatta* Martynov, 1937; *Soliblatta* Lin, 1986; *Taublatta* Martynov, 1937; *Taublattopsis* Vishniakova, 1985; *Thuringoblatta* Kuhn, 1938, and probably some other insufficiently known genera (Vršanský and Ansorge 2007).

Geographic range: cosmopolitan.

Stratigraphic range: P/T–K/Pg.

Diagnosis (after Vršanský 2000): Cockroaches with long external ovipositor. Usually large cockroaches with forewing 15–30 mm (rarely up to 60 mm or under 13 mm). Both wings membraneous or leathery. Forewing with branched SC, RS expanded, M and Cu both richly branched. Cu S-shaped with most branches ending near wing apex. Clavus not surpassing wing midlength. Intercalary veins thick, dark color pattern rather typical (anterior part darker with a pale field). Hind wing with fan-like pleating. SC branched or reticulated, long. R1 and RS differentiated and abundantly branched. M branched. CuA with many secondarily branches and possibly with several blind branches that may also be secondarily branched. Wing usually with many reticulations.

Okras Gen.N.

Type species: *Okras sarko* sp.n. by monotypy.

Differential diagnosis: It is unique within family in small size and distinct dark coloration with pale macula. It is unusual that SC is short and simply dichotomised.

Description: as for species.

Systematical remarks: Distinct coloration is the genus autapomorphy. The family is generally very conservative. It is very probable that the genus evolved from partially colored taxa such as *Nuurcala* Vršanský, 2003.

Derivation of name: *okras* is after *okraska* (Slavic for wonderful lady or coloration). Gender masculine.

Okras sarko sp.n. (Figs. 10a; 15g).

Holotype: 3791/582 (forewing). Assemblage 275/1.

Additional material: 3791/567 (forewing). Assemblage 275/1.

Description: Forewing very wide and short, length/width 10/3.8 mm; with distinct total monochromatic coloration and pale small dot in the radial area. SC short, two branches long. R1 sigmoidally curved, short, with secondarily branches, 14 veins meet margin; RS differentiated, strong, sigmoidally branched, with at least six (probably nine) veins at margin; M rather simplified, with long branches (3); CuA sigmoidally curved, differentiated into two stems (4 + 2). Clavus short, with secondarily densely branched A.

Derivation of name: *sarko* alludes to sarcophagus, from Greek σαρξ—sarx (meat) and φαγειν—fagein (to eat). *Okras sarko* is a palindrome. Gender masculine.

Character of preservation: two complete forewings.

Nuurcala Vršanský, 2003

Type species: *N. popovi* Vršanský, 2003. Böön Tsagaan Nuur.

Composition: *N. obesa* Wang, 2013. Yixian; *N. srneci* Vršanský, 2008. Khurilt.

Stratigraphic range: middle Jurassic (present publication and unpublished data of J. H. Liang from Daohugou)—lower Cretaceous.

Geographic range: Laurasia (Asia).

Diagnosis (after Vršanský 2008): The genus is characterised by its small size and by having both pairs of wings shortened with characteristic coloration comprising dark anterior margin, pale anterior field followed by darker stripe(s).

Nuurcala cela **sp.n.** (Figs. 9a–d; 15b, e).

Holotype: 3791/926. A forewing. Assemblage 275/1.

Paratypes: 3791/588, a forewing; 3791/468, 3776, complete specimens. Assemblage 275/1.

Possibly an additional material: 3791/105±. A forewing fragment. Assemblage 268/4.

Differential diagnosis. Forewing length 14–17 mm, it is somewhat transitional among *N. srneci* (11.5–12.1 mm), *N. obesa* (22 mm) and *N. popovi* (17–26 mm).

Description: Pronotum large, transverse, widest near base, with coloration (obscurely preserved). Forewing elongated, but short (length/width 14–17/6.0 mm). Forewing margins parallel (except for base); SC short (as long as clavus) with few (2–3) branches. R strongly sigmoidal, simply dichotomised, only one branch is secondarily branched, conservatively (unusual) with 15 veins meeting margin. M conservatively with eight veins meeting margin (unusual) curved to both anterior and posterior sides around apex. CuA is expanded and also very conservative in number of branches (8–9). CuP short, a variable, usually dense, with 6–15 veins at margin (some veins are secondarily branched). Coloration contains a pale fenestrum in radial area covered with very dark stripe and less dark wide field. Hindwing with numerous weakly branched CuA and similar number of A2 (18 or more).

Remarks: Conservativism in the number of R and M veins is highly unusual and might represent specimens from the single population (specimens 3791/105 from 208/4 and 3791/294 from 268/4) are too fragmentary to reveal comparative data in this respect; but stratigraphical difference suggests at least populations 208/4 and others were different—specimens might actually represent (an)other species, but at the present state of the knowledge, it is impossible to discriminate it from the present species. The total number of veins meeting margin (ca. 37) is equal to *N. srneci*. Pronotum form is identical (transversely asymmetrical) to *N. obesa* and *N. srneci*, unlike *N. popovi*. The species occurs in at least three very different Bakhar assemblages, although it can not be excluded specimens from assemblages 208/4 and 268/4 represent a separate (sibling) species. Hindwing of the paratype 3791/3776 is very unusual in possessing numerous longitudinal CuA and A2 and no other hindwing from the assemblages 275/1 and 208/4 correspond to this type of hind wing.

Derivation of name: *cela* is Slavic for cell and at the same time the entire (complete as *celá*). Gender feminine.

Character of preservation: two complete specimens, three forewings.

Caloblattina **Handlirsch, 1906**

Type species: *Caloblattina mathildae* (Geinitz, 1883).
Composition: *Caloblattina liasina* (Giebel, 1856), *Caloblattina rubens* Vršanský, 2003.
Stratigraphic range: Rhaetian—Albian.
Geographic range: Laurasia (Germany, England and Mongolia).
Diagnosis (after Vršanský 2003): Large species with rather wide costal field and generally wider wings than *Liadoblattina* and *Rhipidoblatta.*

Caloblattina vremeni **sp.n.** (Figs. 2a; 10fh; 14aa; 15m, s).

Holotype: 3791/496. Assemblage 275/1.
Paratypes: 418±, 422±, 424±c, 431±, 437, 439±, 440±, 445, 448, 455, 462, 469, 479, 508, 509, 547, 571c, 584, 585, 539c, 573, 58?p (number damaged), 601±c, 604±, 607±, 609±c, 644, 651, 679, 680b, 685, 702, 723, 739 celý±, 740±, 741, 756, 777, 807(803?), 826p, 847, 850c, 862, 690, 896ff, 900, 942; 899(889?) (hindwings). All assemblage 275/1.
Type horizon: Assemblage 275/1.
Differential diagnosis: Differs from *C. liassina* (14.5 mm long) in being larger, from *C. rubens* (about 30 mm) and *C. mathildae* (23–30 mm) in being smaller (18–22 mm).
Description: Forewing margins more or less parallel, apex sharp. Forewing length/width 18–22/5.2–9 mm. Costal area wide, SC branched, with three or more veins meeting margin. Veins numerous, R1 (11) and RS (3) differentiated, stem R sigmoidal, radial field not meeting apex. M expanded, with about 11 veins meeting marging, posteriormost veins curved posteriorly. CuA extremely expanded, with eight veins at margin, nearly reaches apex, posteriormost veins sigmoidal. CuP simple, clavus smooth, A branched, not numerous. Hindwing shorter than forewing, 19–22 mm long.
Systematic remarks: Size difference in these extremely size-variating cockroaches, especially from *C. mathildae* is not significant, but along with significant time gaps (Early Jurassic vs. Middle-Late Jurassic), the erection of a new species appears safe.
Derivation of name: *vremeni* means Slavic for "of the time" or "in the belt". Gender feminine.
Character of preservation: 49 specimens, 11 hindwings, one pronotum, four clavi.

Rhipidoblattina **Handlirsch, 1906**

Type species: *Mesoblattina geikiei* Scudder, 1886, p. 454.
Compostion: *R. beipiaoensis* (Hong, 1983), Haifanggou, Liaoning Province, China, the middle Jurassic; *R. boya* Martin, 2010, Mintaja, Australia, lower Jurassic; *R. bucklandi* Handlirsh, 1906–1908, England, UK, J/K; *R. chichengensis* Hong,

1997, Chicheng Country, China; *R. decoris* Lin, 1978; *R. emacerata* Zhang, 1986, Hebei Province, China, Jurassic; *R. forticrusa* Lin, 1986, South China; *R. fuxinensis* Lin, 1978, Liaoning Province, China, Lower Cretaceous; *R. geikiei* Handlirsch, 1906–1908, Browns Wood, England, UK, the Lower Jurassic; *R. gurvaniensis* Vishniakova, 1986, Mongolia, Lower Cretaceous; *R. hebeiensis* Hong, 1980, Hebei Province, China, the Middle Jurassic; *R. jilinensis* Lin, 1994, Jiutai, Jilin Province, China, Lower Cretaceous; *R. kisylkiensis* Martynova, 1951, Kyzyl-Kiya, Russia, Lower Jurassic; *R. lanceolata* Hong, 1980, Chengde, China, Middle Jurassic; *R. laternoforma* Lin, 1978, Chaomidianzi, Liaoning Province, China, Lower Cretaceous; *R. liaoningensis* Hong, 1980, Chaoyang, Liaoning Province, China, Lower Jurassic; *R. liugouensis* Hong, 1983, Xiaofanzhangzi, Hebei Province, China, Middle Jurassic; *?Rhipidoblattina lonchopteris* Hong, 1980; *R. longa* Hong, 1980, Chengde Basin, China, Middle Jurassic; *R. maculata* Vishniakova, 1968, Karatau-Mikhailovka, Kazakhstan, Upper Jurassic; *R. radipinguis* Lin, 1986; Jiuquan Basin, Gansu Province, China, Mesozoic; *R. tenuis* Hong, 1983, the Middle Jurassic; *R. tulunensis* (Vishniakova, 1983), Vladimirovka village, Irkutsk Region, Russia, Lower Jurassic; *R. yanqingensis* Hong, 1997, China, Upper Jurassic; *R. lacunata* Barna, 2014. Chernovskie Kopi, Transbaikalian Russia, J/K (Characters of *R. spathulata* Hong, 1982, suggest it belongs to Raphidiomimidae.)

Stratigraphic range: Lower Jurassic—Upper Cretaceous.

Geographic range: cosmopolitan.

Original diagnosis (after Handlirsch 1906–1908): Well-developed intercalaries and distinct cross-veins; M and Cu branches oriented to the apical outer margin of wing is also characteristic feature for *Lithoblattina, Malmoblattina* and *Elisama*. Body of moderate size; forewings weakly sclerotised, exceeding the length of abdomen. Forewing anterior margin straight or weakly convex, posterior margin straight, apex weakly narrowed, placed approximately symmetrical along the longitudinal axis of forewing; length to width ratio of forewing 3.3–3.9: 1; SC ending at the level of A area, weakly dichotoming; R weakly curved, almost reaching forewing apex, occupying nearly 1/2 of its width; M richly dichotoming distally from CuA, with comb-like distributed branches directed to the outer margin; CuA reaching apex of posterior margin, branched into two stems; CuP evenly arcuate; A area high, elongated, with its length to width ratio 2.5: 1, A1 simple, A2 richly branched, majority of its branches distally dichotomising (modified after Vishniakova 1968).

Rhipidoblattina bakharensis **sp.n.** (Figs. 10e; 14y; 15k).

Holotype: 3791/574. Assemblage 275/1.

Paratypes: 3791/465, 480c, 522, 523, 541, 576, 579, 610±, 622, 660, 734, 745±, 821, 843, 949 (forewings); 3791/425, 485, 590, 593, 828 (hindwings). All assemblage 275/1.

Differential diagnosis: Most species differ in size (see Barna 2014). The same size *R. emacerata* (15/4 mm); *R. fuxinensis* (14/6 mm) and *R. spathulata* (15/4 mm) differ in wing proportions. Most similar *R. jilinensis* (13/4.5 mm) and *R. chichengensis* (14/5 mm), both with identical forewing proportions differ in having larger clavus.

Description: Forewing rather short; length/width 13.5–14/5 mm. SC short, with three branches meeting margin. R (15) field short and wide, covering nearly a half of the wing, RS differentiated, most R veins simply dichotomised. M standardly developed and reaching apex, with seven veins at margin. CuA expanded, with nine veins at margin, CuP fluent; A with 11 veins at margin (branches secondarily branched); diagonal kink present.

Systematic remarks: All species within the genus are highly similar. Specimens of some other species, if present in the same locality would be impossible to discern. Nevertheless, most of the species differ in size and/or slightly in wing proportions. The most similar species, *R. jilinensis* and *R. chichengensis* have a significant time period gap (Lower Cretaceous and Upper Jurassic, respectively), and thus, erection of a new species (which is not self-purposive) is safe. Unfortunately, no more details area available that would extend beyond the species variability of these cockroaches and conspecifics cannot be definitely excluded.

Derivation of name: after Bakhar locality. Gender feminine.

Character of preservation: forewing: 16(1c), hindwings: 5.

Rhipidoblattina sisnerahkab **sp.n.** (Figs. 2g; 8b, d; 10b; 14ac; 15a, d).

Holotype: 3791/ 357 = 270. All wings with pronotum. Assemblage 268/14.

Paratypes: 3791/187, 336, 344c, 346, 361, 364, 367, 369 (forewings); 173, 180, 194, 199, 211, 213, 221, 224, 230, 236, 238, 245, 247, 253, 260, 287, 295, 299, 310, 324, 329, 337, 338, 341, 351±, 352, 356, 986, 996, 999, 1052, 1044 (hindwings); assemblage 268/14.

Additional material: 3791/165, 166, 168, 172, 174±, 178 ± ffhh, 179±, 195, 197, 205, 200, 208, 215, 217, 227, 233, 234, 241, 258, 259, 262, 277, 279, 286c, 290, 346, 394, 269, 273c a7, 325, 327, 329fh, 345, 348, 363c, 965±, 968, 969±, 370, 978, 979, 980±, 994, 1006, 1010, 1011c, 1012, 1015c, 1023, 1024c, 1029, 1030, 1134–1135, 1137, 1038c, 1041, 1046, 1047, 1051, 1058, 1059, 1078, 1079, 1089, 1095, 1118, 1129, 1130, 1131, 1033, 1136, 1144, 1150, 1158, 2024, 5052, 5053, 5146, 5148–5149, 5159, 5163, 5164, 5165, 5181c, 5199, 5204c (forewings); 3791/110, 166, 169±, 170, 173, 180±, 183, 184, 188, 198, 250, 293, 204, 209, 223, 232, 255, 256, 275, 292, 297, 305, 314 14 mm, 343, 353, 357–358, 376, 964±, 968±, 983±, 986, 987±, 1007, 1014, 1034, 1036, 1037, 1045 **2x**, 1048, 1054, 1067, 1071, 1087, 1091, 1094, 1097, 1098, 1100, 1106, 1107, 1111, 1115, 1117, 1120, 1125, 1151, 1140, 1143, 5144±, 5150, 5186, 5157, 5190, 5202 (hindwings); assemblage 268/14.

Additional material: 3791/111, 115, 116, 119, 120, 122 (forewings), 3791/121 (hindwing); assemblage 268/4. 3791/133, 140c 6.5 mm (forewings); assemblage 268/8.

Additional material: 3791/144±, 148, 152, 153, 155, 157, 159, 158, 160 (forewings); 3791/145±, 146±, 147, 151, 154, 156 (hindwings) (268/12).

Additional material: 3791/375, 379, 384, 387, 389, 391, 396, 401, 411 (forewings); 3791/380, 381, 382, 383, 386, 391, 395, 399, 400, 407, 409–410, 414 (hindwings) (268/19).

Differential diagnosis: Differs from *R. bakharensis* in averagely smaller size (forewing length 11–14, exceptionally up to 16 mm compared with 13.5–14 mm). Most similar species are slightly smaller *R. bucklandi* (11 mm) and identical in size *R. radipinguis* (12 mm) from China. For further detail see remarks.

Description: Pronotum transverse (4/4.5 mm), widest in the central part. Small posterior extension present. Coloration sophisticated, with pale central, basal and lateral areas (see Figs.). Forewings moderately elongate (11–16/4–4.5 mm) with apex sharpened anteriorly. SC as long as clavus with 2–3 veins meeting margin. Radial field wide nearly reaching or reaching the half of the wing width. R undifferentiated, with 14–16 veins at margin, some veins tertiary dichotomised. M rather simplified, with 4–7 veins meeting margin. CuA expanded, nearly reaching apex, with about eight veins at margin. Two simple anterior anal branches preserved in both sides. Hindwing shorter than forewing (11–13 mm) with round apex. SC very long reaching the half of wings, branched, with three terminal branches. R1 (6) and RS (6) differentiated. M standard, with five nearly straight veins meeting margin. First dichotomy near middle of the wing. CuA simply branched, veins slightly curved (6), CuP simple.

Systematic remarks: It is impossible to formally discriminate specimens from associated bed (Assemblage 268/12), but it is possible that these specimens represent a closely related, separated distinct sister species. In spite of size difference, sister species is probably *R. bakharensis*. *R. bucklandi* was recorded in a very differernt time period (J/K). Age of *R. radipinguis* is unknown, but that species originates from China (Upper Jurassic or Early Cretaceous) and thus as geographically different can be regarded for a separates species with a high degree of confidentiality (more safe that categorisation within this species).

Hindwing specimen 3791/302 reveals a structure similar to breaking sutura of termites, and thus, it cannot be excluded hidwings were detached during some lifecycle stage. Nevertheless, the preservation state disallows explicit statement.

All characters are in-generic, i.e. no basal or derived states were observed allowing for systematic hierarchical classification of the species within the genus or clarifying the position of the genus within the system. Only pronotum coloration with pale central area and shaped dark macula forming lateral and basal pale areas can be considered for (ecologically) advanced.

Derivation of name: *sisnerahkab* is reversed *bakharensis*. Gender feminine.

Character of preservation: 97 forewings (1ffp, 1ffh, 1fh, 10c), 98 hindwings (268/14); six forewings, one hindwing (268/4); two forewings (1 clavus) (268/8); eight forewings, seven hindwings (268/12); nine forewings, 13 hindwings (268/19); Total 244 (122 forewings, 119 hindwings).

Rhipidoblattina konserva **sp.n.** (Figs. 1f; 8c; 9f; 10c, g).

Holotype: 3791/278 22/. Assemblage 268/14.

Paratypes: 3791/162±, 163±, 164±, 167±, 175±, 181±, 185±, 190, 191±, 193, 196c, 201, 206c, 214c, 219, 222c, 239c, 240c, 244, 251, 254, 261, 263, 272, 274c, 281, 289, 291, 293, 296c, 298, 300, 311c, 312, 318, 340c, 349, 350c, 355c, 359, 363, 962±, 972±, 976, 977±, 959 (complete specimen), 960±, 973c±, 982,

989±, 995c, 997, 1001, 1002, 1008, 1009, 1017, 1019, 1022, 1031, 1049c, 1050, 1055±, 1056, 1061, 1062, 1067, 1068, 1070, 1076, 1077, 1081, 1083, 1084, 1085, 1090, 1099, 1101, 1103, 1108, 1109c, 1112, 1116, 1122c, 1132, 1133, 1149, 1152c, 1154, 1155, 1157, 1159, 1160, 1162, 1164c, 1165, 1169, 1172c, 1173, 2038, 5142, 5143, 5145±, 5151, 5153, 5161, 5167c, 5168, 5169, 5170, 5174c, 5175, 5176, 5178, 5179fh, 5187, 5191, 5193, 5195, 5198, 5201c, 5203c (forewings); 3791/166, 170, 171±, 176±, 177±, 182, 184, 189, 207, 210, 212, 216, 218, 220, 228, 229, 235, 249, 252, 266, 267, 271, 282–283, 309, 315 = 285, 317, 319, 322, 342, 347, 360, 963±, 967±, 971±, 961±, 966, 974±, 981±, 991, 993, 998, 1004, 1005, 1013, 1016, 1020, 1035, 1039, 1045, 1063, 1064, 1069, 1073, 1072, 1080, 1086, 1093, 1096, 1104, 1105, 1119, 1121, 1124, 1145, 1153, 1161, 1163, 5166, 1168, 1171, 5156, 5158, 5160, 5162, 5172, 5173, 5180, 5184, 5185 (hindwings); 3791/984±, 1126 = 1140, 1166 (pronota). All assemblage 268/14. Number 3791/177 is given to two specimens (one belongs to Blattulidae).

Additional material: 3791/114. Assemblage 268/4.

Additional material: 3791/127±, 125±, 137c, 138, 139c, unmarked (forewings); 3791/136 (hindwing). All assemblage 268/8.

Additional material: 3791/142 ± (forewing); 3791/143±, 150 (hindwings). All assemblage 268/12.

Additional material: 3791/371±, 373c, 376, 377, 378, 397, 403, 406, 412 (forewings); 3791/372±, 385, 390, 404, 413 (hindwings). All 268/19.

Type horizon: Assemblage 268/14.

Differential diagnosis: Larger than any representative of the genus.

Description: Forewing without coloration, 25 mm or longer and 6–7 mm wide. Radial field covers half width of the wings, R strongly sigmoidal, 11 vein preserved at the margin, total number of veins much higher due to partial preservation of wing. M and CuA similarly expanded, together with at least 15 veins meeting margin. A with 11 veins preserved.

Systematic remarks: It is impossible to formally discriminate specimens from associated bed (Assemblage 268/4).

Derivation of name: *konserva* is Slavic for "the can". Gender feminine.

Character of preservation: seven forewings (2 clavi), one hindwing (268/8); one forewing (268/4); one forewing, two hindwings (268/12); nine forewings (one clavus), five hindwings (268/19); one complete specimen, 121 forewings (25 clavi, one forewing with hindwing), 80 hindwings, three pronota (268/14); totally one complete specimen, 139 forewings (28 clavi, one forewing with hindwing), 88 hindwings, three pronota.

Solemnia Vršanský, 2008

Type species: *Solemnia alexandri* Vršanský, 2008.

Stratigraphic range: (J1, J2) J3.

Geographic range: indigenous to Mongolia and Inner Mongolia, China.

Diagnosis (after Vršanský 2008): Forewing extremely long and narrow, with distinct coloration; SC extremely short, simple or 2(–3) branched; R almost straight, with most branches simple; M reduced to several long, straight veins; CuA expanded, with the posteriormost branch richly comb-like branched; anal veins branched. Hindwing with differentiated, richly branched RS, simplified M and secondarily branched CuA. Head globular, body and legs slender.

Solemnia togokhudukhensis sp.n. (Figs. 1e; 2e; 8a; 9g, h)

Holotype: 3791/456. A forewing fragment.

Paratypes: 3791/428±, 441c±, 446 ± ff, 451, 482, 492, 511c, 578, 591c, 596±, 600c, 602±, 619, 634fh, 647, 650, 652, 655c, 658, 665, 672, 705c, 717, 721, 746±, 751, 754c, 825c, 835c, 890c, 908c, 931c (forewings); 744, 616±, 932, 918, 677, 946 (hindwings); 427 ± (pronotum). All assemblage 275/1.

Additional material: 3791/1121(1181?), 1175±, 1176±, 1178±, 1179, 1180±, 1181, 1185, 1191, 1193. All assemblage 275/2.

Type horizon: Assemblage 275/1.

Diferential diagnosis: It differs from the type species (16–18.3/4.3–4.7 mm) and a possible congener *"Rhipidoblattina"hebeiensis* Hong, 1980 (20/6 mm) in size in upper limit and transitional size respectively and also by having wider base of the costal area.

Description. Forewing long 18 mm and 4.7–4.8 mm wide. Costal area narrow, but expanded posteriorly in relation to wing base. SC sigmoidal, long and simple; R with primarily branched stems; M and CuA normally developed; CuP simple, clavus up to 5.2 mm long; A secondarily branched and with a diagonal kink. Coloration on pale membrane formed with diagonal dark and less dark stripes.

Systematic remarks: The species is categorised within this genus on the basis of identical general morphology and also characteristic coloration (this coloration is nevertheless present in other species of the family Caloblattinidae (*Nuurcala*), but also Fuziidae (*Longifuzia pectinata* Liang et al. 2019). This inference is supported by close spatiotemporal relation of the Khoutiin-Khotgor locality within the same Khoutiin-Khotgor complex with Bajan-Ul rock statum of overlaying assemblages 327 and 328. Species is nevertheless different from its sibling species *Solemnia alexandri* Vršanský, 2008, which is discussed below. Relation of the species is very close, no diagnostic characters except for the wider base are recognised and discrimination is mostly based on size. Wider base is considered for plesiomorphic character within genus.

Derivation of name: after Togokhuduk. Gender feminine.

Character of preservation: 33 forewings (1ff, 11c, 1fh); 6 hindwings, 1 pronotum (275/1); 10 forewings (275/2); TOTAL 40 (11c; 6hw; 1p; 1ff; 1fh).

Blattoidea Latreille, 1810
Mesoblattinidae Handlirsch, 1906

Type species: *Mesoblattina protypa* Geinitz, 1880.

Composition: *Actinoblattula, Aporoblattina, Archimesoblatta, Artitocoblatta, Austroblattula, Basiblattina, Beviblattina, Blattidium, Blattidium medium, Breviblattina, Durdlestoneia, Etapia, Fusiblatta, Fusoblatta, Gondwablatta, Hispanoblatta, Hongaya, Ijablatta, Itchetuja, Jingyuanoblatta, Kemerowia, Kulmbachiellon, Laiyangia, Lithoblatta, Malmoblattina, Mesoblatta, Mesoblattellina, Mieroblattina, Mongolblatta, Nipponoblatta, Nogueroblatta, Nymphoblatta, Pareinoblatta, Perlucipecta, Pulchellablatta, Rhaetoblattina, Rithma, Schambeloblattina, Sivis, Soliblatta, Stantoniella, Turoniblatta.*

Stratigraphic range: earliest Early Jurassic—terminal Cretaceous.

Paleogeographic range: cosmopolitan.

Diagnosis (after Vršanský and Ansorge 2007): Cockroaches with parallel forewing margins and regular venation, branched SC, more or less straight R stem and branched M and CuA. CuP not curved sharply, few anal veins branched or simple. Hindwing with simple SC, differentiated R1 and RS, branched straight M and branched CuA. Blind CuA branches might occur. Remigium also contains simple CuP and A1. Ovipositor rudiments might be present.

Perlucipecta Wei et Ren, 2013

Type species:*P. aurea* Wei et Ren, 2013; Yixian, China.

Composition: *P. vrsanskyi* Wei et Ren, 2013; Yixian, China; *P. santanensis* Lee, 2016; Crato, Brazil.

Stratigraphic range: J1-K2.

Paleogeoraphic range: cosmopolitan.

Perlucipecta cosmopolitana sp.n. (Figs. 2c, f; 8f; 14ad).

Holotype: 3791/724 = 731. 5 mm clavus length.

Additional material: 742±, 897 (forewings); 813 (hindwing). All assemblage 275/1.

Type horizon: assemblage 275/1.

Differential diagnosis: It differs from the type species *P. aurea* in having bigger distance among CuP and A1; from *P. vrsanskyi* Wei et Ren, 2013 and *P. santanensis* Lee, 2016 in being larger.

Description: Eight simple anal veins preserved in standardly vaulted clavus, first two of them terminate in CuP, A1 distant from CuP basally.

Systematic remarks: the species can be categorised within this most common Mesozoic genus of the family on the basis of characteristic coloration and simple slightly posteriorly curved anal branches.

Derivation of name: *cosmopolitana* is Latin for cosmopolitan. Gender feminine.

Character of preservation: forewing: 3(1c), hindwing: 1.

Corydioidea Saussure, 1864
Blattulidae Vishniakova, 1982
Blattula **Handlirsch, 1906**
Blattula velika **sp.n.** (Figs. 2 h; 7f; 14 k, af).

Holotype: 3791/416. A forewing with a hindwing and body parts.

Additional material: 379 1/434, 495, 497, 512, 519, 524, 561, 671, 720, 786, 851, 952. All assemblage 275/1.

Type horizon: Assemblage 275/1.

Differential diagnosis: It differs from all species at Bakhar in significantly larger size, forewing up to 13 mm long. *Blattula mongolica*—a sister species has identical form of the forewing and is stratigraphically adjacent in Shar-Teg, but is slightly smaller, 12 mm forewing length.

Description: Major veins, cross-veins and intercalaries distinct. Forewing length/width 12.5–13.0/3.6–4.0 mm position of round apex in nearly central. Costa distinct, long, probably reaching apex. Costal field arcuate, but narrow. SC simple but possibly with terminal veinlets; R stem arcuate basally, then straight, with 14–17 branches at the margin, without secondary bifurcations, RS not differentiated. M with about nine branches, CuA with about four branches. Clavus narrow, up to 5 mm long, with five anal veins. Hindwing probably as long as forewing or shorter. R1 and RS (5 + 5) differentiated, M not reduced, with about five branches.

Systematic remarks: Generally *Blattula* is mostly very conservative in shape, and thus, diagnostics at the species level is very difficult and mostly restricted to size. Thus, errors in the taxonomic procedure cannot be excluded and in the matter of fact sister species like *B. velika* and *B. mongolica* might actually represent different populations of the same species. Nevertheless, size difference is sufficient for formal discrimination, and spatiotemporal difference makes erection of a new species comparatively safe. Size difference 0.5 mm is substantial in this small and actively living species.

Derivation of name: *velika* is Slavic for major. Gender feminine.

Character of preservation: 14 forewings (1 with hw, 1c).

Blattula vulgara **sp.n.** (Figs. 5q, t; 7a–c; 13r; 14f).

Holotype: 3791/419±. A forewing.

Additional material: 3791/421, 449, 461, 563, 470, 487, 506, 535, 566, 586, 597, 615±, 630fhh, 631, 632, 637, 641, 646, 659, 676bff, 673, 678, 712, 728, 738, 748±, 758, 766fh, 798, 809, 814, 820, 834, 837, 841, 852, 869, 872, 885 bfovip, 901, 910, 911, 914, 916, 923, 925, 934, 937 (forewings); 3791/528, 570, 692, 864 (hindwings). All assemblage 275/1.

Type horizon: Assemblage 275/1.

Differential diagnosis: It differs from other representatives of the genus in having wide forewing, short clavus, and non-parallel wing margins. Branched SC is a rare character within family. Posterior CuA with reticulations is autapomorphic.

Description: Forewing 7.1–9/2.2–3 mm long and wide, margins not entirely parallel, apex posed centrally, clavus very short. SC simple or simply branched, sigmoidal, short. Radial field nearly straight, only slightly sigmoidal, very wide, covering half of the wing. R branched, RS differentiated with five veins, R totally amounting to 12–18 veins at margin. M expanded, with 4–6 veins meeting margin. CuA comb-like, short, with 5–7 descending simple veins meeting margin. 5–6 simple anal veins present. Hindwing 7–8 mm long, SC branched, R1 (3) and RS differentiated (7); M (4) and CuA (7) fully developed.

Systematic remarks: The hindwings are categorised within this species on the basis of correlated widening of the radial area, and disproportions, particularly of RS.

Derivation of name: *vulgaris* is Latin for folk. Gender feminine.

Character of preservation: 52 (1 ffh, 1 bff, 1fh, 4h, 1 bfovip).

***Blattula mini* sp.n.** (Figs. 7n; 14q).

Holotype: 3791/516. A forewing.

Paratypes: 3791/452, 478, 481, 484, 486, 490, 525, 536, 532, 555, 558, 603±, 628, 629, 633, 643, 670, 686, 693(683?), 704, 709, 726, 732, 759, 763, 779, 788, 791, 803, 808, 817, 840, 854, 859, 904, 913, 3133 (forewings); 3791/483 (hindwing). All assemblage 275/1.

Additional material: 3791/1174B±. Assemblage 275/2.

Type horizon: assemblage 275/1.

Differential diagnosis: This is the most characteristic appearance of the genus, known by dozens of species throughout the Mesozoic. It differs from all others like most similar (also in size) *B. vidlickai* Vršanský, 2004 and *B. brevicaudata* Vishniakova, 1968, from spatiotemporary adjacent Shar-Teg and Karabastau, respectively, in having centrally posed, contrary to usual, shifted apically apex. *B. langfeldti slightly larger* has sharpened apex.

Description: Forewing narrow (1.3–1.9 mm) with length 4.5–5.6 mm, apex round and slightly sharpened, posed centrally. Main veins and intercalaries thick, cross-veins present. Costal area very narrow, SC very long and slightly sigmoidal, simple. R slightly sigmoidal, nearly reaching apex, with 10–14 veins meeting margin, branches primarily dichotomed, RS indistinct. M conservatively with five slightly vaulted branches. CuA with 4–6 simple veins at margin, CuP simple, very sharply curved, clavus very short. Five or six simple curved anal veins present. Hindwing 5.1 mm long.

Systematic remarks: The taxon is categorised within *Blattula* on the basis of standard blattulid morphology (forewing with oval regular shape, regular venation, regular size of clavus, regular distance among veins, IC present, CW not especially expressed), lack of coloration characteristical for advanced genera other than *Blattula* and common Mesozoic occurrence.

Derivation of name: *mini* is for small. Gender feminine.

Character of preservation: 36 forewings (one with body, one with hindwing, two isolated clavi); one hindwing.

***Blattula mikro* sp.n.** (Figs. 7m; 14p).

Holotype: 3791/5051. A forewing.

Type horizon: assemblage 275/1.

Differential diagnosis: Differing from all Blattulidae except its sister species *B. flamma* in being very wide, and from its sister species *B. mini* in having longer SC.

Description: Forewing wide (1.8 mm) with length 4.8 mm, apex round, located centrally. Main veins thick, intercalaries distinct, cross-veins likely present, preserved as membrane disturbances. Costal area standard, SC very long and sigmoidal, simple. R strongly sigmoidal, nearly reaching apex, with 11 veins meeting margin, branches primarily dichotomed, RS indistinct. M reduced to few (3) slightly vaulted long branches, similar in character to CuA. CuP simple, very sharply curved, clavus extremely short, reaching only about quarter of wings length. Five or six simple curved anal veins present.

Systematic remarks: The taxon is categorised within *Blattula* on the basis of standard morphology, lack of characteristic coloration and common Mesozoic occurrence. *B. mini* from the same horizon is significantly more elongated and have narrower costal field than *M. micro*.

Derivation of name: *mikron* is Greek for small. Gender feminine.

Character of preservation: 1 complete forewing.

Assemblage 268/4

The assemblage representing a bed within 268 packages.
Unidentifiable cockroaches are: 3791/112, 118, 124, 113.

Assemblage 268/8

The assemblage representing another distinct bed within 268 packages.
Unidentifiable cockroaches are: 3791/128, 132, 134, 135, 126±, 130, 141p.

A unique small species with forewing length/width 9/2.2 mm species of Raphidiomimidae (3791/131) also occurs (marked as Raphidiomimidae sp. in Table 5).

Assemblage 268/12

The assemblage representing another bed within 268 packages.
Unidentifiable cockroaches are: 3791/161 (hindwing); 3791/149 (unidentifiable).

Assemblage 268/14

Unidentifiable cockroaches are: 3791/180h, 186c, 225c, 226, 202, 328, 255, 288, 323, 332, 339, 280, 301, 307, 330, 331, 362, 978, 5155, 1000 (forewings); 3791/183, 284, 316, 320, 237, 366, 335, 5182, 5189, 5154, 5171, 5183, 5177, 990, 1018, 985±, 1021, 1003, 1027, 1075, 1082, 1092, 1146, (hindwings); 3791/304 (immature individual); 3791/169, 203, 326, 333, 264, 313, 354, 970±, 1060, 1074, 1042, 1053, 1102, 1122–1123, 1156, 1170(1179), 1127, 5152, 5194, 5188, 5196, 5197 = 5200 (unidentifiable).

3791/992, 242, 248, 257, 231, 303, 1025, 5192, 975 ± p, 1028, 1032, 1040, 1066p, 1088, 1114, 1138, 1139, 1167, 1141, does not belong to Blattaria or has been damaged during the deposition.

This assemblage is represented only by large specimens, weak preservation optically causes selection of larger and harder forewings. In oryctocenoses, other than formalised herein species were likely present. This assemblage contains *Hra disko* ($n = 5$; one more or less complete specimen and two hidwings), which suggest a low transportation and shallow water not causing disarticulation.

Assemblage 268/19

Unidentifiable: 3791/392, 375c, (forewings); 3791/394, 402, (hindwings); 3791/408, 398, 383, 5069 (unidentifiable).

3791/388 does not belongs to Blattaria.

Liberiblattinidae Vršanský, 2002
Dostavba Gen.N.

Type species: *Dostavba pre* sp.n. by monotypy.

Differential diagnosis: It differs from all representatives of the family in rather short SC basally branched. New genus differs from related *Stavba* Vršanská et Vršanský, 2019 in elongate forewing and in having sharp apex posed centrally. *Spongistoma* Hinkelman in Sendi et al. (2020b) has more extensively branhced R.

Description: as for species.

Systematical remarks: Elongate forewing is a synapomorphy of new genus and *Stavba*-group—a rich complex of genera known from Myanmar amber. Basally branched SC is a strong symplesiomorphy at the level of the family of so-called *Voltziablatta*–group represented also by "Triassoblattidae".

Derivation of name: *Dostavba* is composed of *do* (Slavic for prior) and *stavba* (Slavic for building), related to relation with genus *Stavba*. Gender feminine.

Dostavba pre **sp.n.** (Fig. 14c).

Holotype: 3791/405. A forewing.

Type horizon: assemblage 268/19.

Description: Forewing elongate, 8/1.9 mm, with sharp apex posed centrally. SC rather short, basally branched, with four veins meeting margin. R nearly straight, RS undifferentiated, R1 branched, totally R with 13 mostly simple veins meeting margin; M simplified, nearly straight, with six veins at margin; CuA slightly sigmoidally curved, with six veins meeting margin, reaching apical part; CuP elongate, clavus long, A branched.

Derivation of name: *pre* is Slavic for prior. Gender feminine.

Character of preservation: 1 complete forewing.

Assemblage 208/2

Unidentifiable: 3791/55, 54, 61, 48, 80, 83, 75, 84; 87p (5.0/4.6 mm).

It is notable that *Vrtula sama* Vršanský, 2008 ($n = 24$) similar to *Truhla vekov* (Bakhar) is the only known cockroach from the respective locality, Shin-Khuduk (Lower Cretaceous), so this record might relate to the early succession series of the present horizon. It might be supported by only few taxa found here at 208/2. *Blattula universala* and particularly *Raphidiomima chimnata*, although representing entirely different, predatory taxon have a similar, very unusual elongated habitus.

Also, assemblages 208/3 and 203/6 likely represent the same biom.

Blattulidae Vishniakova, 1982
Blattula Handlirsch, 1906
Blattula flamma **sp.n.** (Figs. 7e, k; 14h, m, o).

Holotype: 3791/64.

Additional material: 3791/50±, 56, 58, 59, 68, 73, 76, 89, 90, 109, 505 (all assemblage 208/2); unlabelled specimen (assemblage 203/6);

Type horizon: assemblage 208/2.

Differential diagnosis: It differs from roughly coeval and highly variable *B. extensa* Vishniakova, 1982 (forewing length = 3–3.5 mm) and *B. iensis* Vishniakova, 1982 (forewing length = 8 mm) Iya from Russia in size, which seems to conserved around 6.5 mm in the present taxon. Extremely similar, practically identical to 3791/109 is *Blattula mongolica* Vršanský, 2004 from Shar-Teg in Mongolia more significantly differing in size (12 mm).

Description: Forewing wide, 6.0–7.0 mm long and 2–2.1 mm wide. Costal field very narrow, C strong, SC very short, nearly straight, simple. Radial field short and wide, main stem of R strongly sigmoidally curved, R veins branched and up to 13 meeting margin; M main stem curved, simplified, with about 3–4 long veins meeting margin; CuA with about 3 very long, but curved simple branches; CuP simple, sharply curved, clavus short. 5–6 curved simple A.

Systematic remarks: The present species is most closely related to the middle Jurassic representatives of the genus mentioned above (Iya, Russia), rather than to common Late Jurassic and Cretaceous species, which has more conservative and advanced shape (for revision see Barna 2014).

Derivation of name: *flamma* is Latin for flame. Gender feminine.

Character of preservation: 13 forewing fragments.

Blattula universala **sp.n.** (Figs. 7d; 14g).

Holotype: 3791/88. A forewing.
Paratype: 3791/71; assemblage 208/2.
Type horizon: assemblage 208/2.

Differential diagnosis: It differs from other representatives of the genus in having forewing apex posted centrally and elongated clavus.

Description: Forewing without coloration, significantly elongated, 9 mm long and 2.2 mm wide, apex sharp, located centrally, but wing decease more fluently in posterior side. Cross-vein invisible; intercalaries distinct only locally, especially among R and M and in terminal part. Costal field narrow, costa strong and distinct; SC sigmoidal, short and simple. Radial field comparatively standard (it is rather narrow, but all wing is narrow so the proportion remains standard), sigmoidally significantly curved, not reaching apex, with 11 veins meeting margin. M expanded, with five veins. CuA with four veins, CuP simple, clavus elongated, with six simple anal veins.

Derivation of name: *universalis* is Latin for universal. Referring to a common appearance. Gender feminine.

Character of preservation: two nearly complete forewings.

Blattula anuniversala **sp.n.** (Figs. 8e; 14b).

Holotype: 3791/645. Both forewings with a hindwing.
Additional material: 3791/623. Assemblage 275/1.
Type horizon: assemblage 275/1.
Differential diagnosis: It differs from its sister species, closely related *B. universala* in blind branches of A connected to CuP.

Description: Forewing 10–12.5 mm long, with anterior proximal margin extended, costal area narrow. SC simple or simply branched, sigmoidal. R sigmoidally curved, frequently branched, with 12–14 veins meeting margin. M strongly sigmoidally curved, reduced to 3–4 veins at margin. CuA expanded, with 6–7 veins at margin. CuP elongate, six main A simple, but with frequent fusions (A-A, A-CuP).

Systematical remarks: Vein fusion connecting posterioromost M and anterior-most CuA is not a taxonomic character and probably represent a deformity (Vršanský 2005; Vršanský et al. 2017), which is present in both wings. Unusual anal veins termi-nated regularly at posterior margin (not running parallel to the margin) allow for categorisation of species in close relation to *B. universala*.

Derivation of name: *anuniversalis* is for non-universal. Referring to a common appearance, but differing from *universala*. Gender feminine.

Character of preservation: two nearly complete forewings (1FFH).

Truhla **Gen.N.**

Type species: *Truhla vekov* sp.n.; and by monotypy.
Differential diagnosis: It differs from all representatives of the family except *Vrtula* Vršanský, 2008, in being extremely elongated and having sharp apex, and from *Vrtula* in having extremely narrow radial field.

Description: as for species.

Systematical remarks: it can be categorised within Blattulidae on the basis of straight simple SC and unspecified RS. The taxon shares the elongated habitus and sharp apex with *Vrtula* and lacks coloration as Jurassic genera such as *Blattula*. The wings of Cretaceous representatives of the family usually have wing colored.

Derivation of name: *truhla* is Slavic for coffin. Gender feminine.

Truhla vekov **sp.n.** (Figs. 7g, i; 14d, n).

Holotype: 3791/52. Anterior half of the forewing (specimen damaged during excavation, not original preservation state).

Additional material: 3791/93, 95, 99, 102 (forewings); 3791/101 (hindwing). All assemblage 208/3.

Type horizon: Bakhar Bed 208/2.

Description: Forewing extremely elongated, apex sharp, posed anteriorly, 8.7 mm long. SC simple, nearly straight, narrow. Radial field extremely narrow, R with 13 veins at margin, three branches dichotomised, others simple. M with three extremely long branches meeting margin. CuA present.

Derivation of name: *vekov* is Slavic for ages. Gender feminine.

Character of preservation: five forewings, one hindwing.

Polliciblattula **Gen.N.**

Type species: *Polliciblattula analis* sp.n.

Differential diagnosis: It differs from all known taxa except *Blattula dubia* Handlirsh, 1939, in being significantly smaller and with miniaturisation adaptations related to reductions. *B. dubia* differs in having standard regular (advanced) wing shape. The wing of *Polliciblattula* shortened with round apex located posteriorly. Hindwing also with extremely reduced venation, while A1 branched.

Description: as for species.

Derivation of name: *pollicis* is Latin for very small (or a thumb—referring both to the shape and size of specimens), combined here with the type family genus *Blattula*. Gender feminine.

Polliciblattula analis **sp.n.** (Figs. 7p; 14i).

Holotype: 3791/92. A forewing.

Type horizon: Assemblage 208/2.

Differential diagnosis: It differs from *P. tatosanerata* in having more expanded CuA and standard M.

Description: forewing very short and wide (length/width 4/1.4 mm), apex round, located posteriorly. Costal field narrow, costa indistinct, SC very short, not reaching third of the wing, straight, simple. Radial field wide with main stem nearly straight, R branches simple or dichotomised, with seven veins meeting margin; M simplified to two sigmoidal long veins of which the anterior one terminally forked and covering apex. CuA with four veins, CuP sharply curved, clavus very short with three curved simple anal veins.

Derivation of name: *analis* is after anal field of the wing which is affected with miniaturisation even more and is comparatively even smaller than the specimen itself. Gender feminine.

Character of preservation: one forewing.

Polliciblattula tatosanerata **sp.n.**

Holotype: 3791/306. A forewing.
Type horizon: Assemblage 268/14.
Differential diagnosis: It differs from *P. analis* in shortened CuA and in having more expanded CuA and comb-like tending M.
Description: Very wide forewing ca. 5 mm long, with apex posed posteriorly. Subcosta nearly straight, RS differentiated with five veins meeting margin, R straight, totally with 13 veins at margin, two of R veins dichotomised. M with five veins at margin, with anterior offshots. CuA with five veins at margin, veins short. CuP sharply curved, clavus short.
Derivation of name: *tatosanerata* is Slavic for "this is not counted". Gender feminine.
Character of preservation: one forewing.

Polliciblattula vana **sp.n.** (Figs. 5p; 13q).

Holotype: 3791/606±. Both forewings and a hindwing.
Type horizon: Assemblage 275/1.
Differential diagnosis: The hindwing is categorised within this genus on the basis of extremely simplified venation unknown in any Mesozoic taxon. Categorisation within the same species is nevertheless unlikely as there are no shared taxa among these assemblages. Anyway, the taxon at least at the genus level apparently passed boundary among 208/2 and 268/14 where an additional specimen within this genus was preserved.

Forewing with extremely simplified M.

Description: hindwing with round apex, with extremely reduced venation. SC simple, extremley short and nearly straight. A significant free area occurs among SC and R1 (2); RS is not reaching apex, with four branches. This area developed some vein distance irregularity due to miniaturisation and vein replacement by fully developed intercalaries (known in Nocticolidae—see Sendi et al. 2020). M simple and straight. CuA with four simple veins at margin, CuP simple. A1 fully developed and even with three blind branches, two basally one centrally. Vannus not preserved.

Forewing very small (4.5 mm long) and proportionaly wide (1.7 mm). Costal area very short and narrow, with simple SC. R sigmoidally curved, reaching centrally posed round apex, with seven veins. M simplified possibly with a single branch; CuA with four more or less straight branches; CuP sharp, clavus short and wide, with four simple anal veins. Intercalaries strong.

Systematic remark: The associated forewings most probably represent the same individual as there are no other such small species in the assemblage.

Derivation of name: *vana* is after vannus which is affected with miniaturisation even more and is comparatively even smaller than the specimen itself. It also means insignificant. Gender feminine.

Character of preservation: one hindwing.

Caloblattinidoidea Vršanský et Ansorge in Vršanský (2000)
Raphidiomimidae Vishniakova, 1973
Raphidiomima **Vishniakova, 1973**
Raphidiomima chimnata **sp.n.** (Figs. 1g; 2b; 6g; 10d, i; 14ab; 15h, p, r).

Holotype: 3791/70. A forewing.

Paratypes: 3791/49±, 51±, 53, 54, 60, 66, 63, 65c, 67, 72, 74 = 85, 77, 78, 79, 81c, 82c, 86 (forewings); 3791/2 (hindwings); 3791/69 (pronotum). All assemblage 208/2.

Type horizon: assemblage 208/2.

Differential diagnosis: The new species differs from the type species *R. chimaera* by simpler shape of the wing (apex of *R. chimaera* shifts to anterior margin and fluently descends posteriorly) and having CW in forewing, and differs from *R. cognata* having significantly smaller forewings (*R. cognata* forewing length over 20 mm, like type species). The coloration of pronotum is like *Graciliblatta bella*, which differs in expanded venation.

Description: Pronotum elongated, widest in the apical third, vaulted, with two longitudinal stripes. Forewing (description based on only well-preserved holotype) significantly elongated, 16–17.7 mm long and 4–5 mm wide. Shape simple, apex sharpened medially, diagonal kink present. Membrane possibly transparent, maculation similar to that of *Cameloblatta*. One distinct small macula present at the top of the clavus. Intercalaries and cross-veins distinct. Costa not distinct, costal area extremely narrow. SC branched, with three and possibly more veinlets. R slightly curved, R1 and RS differentiated with 13 and eight veins at margin, respectively. R1 branches dichotomised. M not reaching apex, sigmoidal, with 12 veins meeting margin. CuA with two main stems and six veins meeting margin. CuA simple, very sharply curved, clavus very short. Nine A veins meeting margin, dichotomised.

Systematical remarks: the present taxon can be categorised within the genus *Raphidiomima* on the basis of extremely elongated forewing, although the coloration is rather similar to much shorter *Cameloblatta* Vishniakova, 1973. *Divocina* Liang, Vršanský et Ren, 2012, is somewhat similar in wing shape, but not extremely elongated and possess nearly uniform distinct coloration. *Fortiblatta* Liang, Vršanský et Ren, 2009 and *Graciliblatta* Liang, Huang et Ren, 2012, have the very primitive venation and differ also in being very wide. *Falcatusiblatta* Liang, Ren et Shih, 2018, is not elongated.

In this species, quite frequent preservation of isolated clavi is present.

Derivation of name: *chimnata* is a stochastical combination of letters, combined from *chimaera* and *cognata*, sister species. Gender feminine.

Character of preservation: 17 forewings (3 clavi), one hindwings, one pronotum.

Raphidiomima krajka **sp.n.** (Figs. 2d; 14z).

Holotype: 3791/776 (clavus). Assemblage 275/1.

Differential diagnosis: The new species differs from the type species *R. chimaera* and other species in having simpler colored anal branches.

Description: Anal veins wide and dark colored, meet margin, dichotomised. Clavus ca. 3 mm long and 1 mm wide.

Systematical remarks: An isolated clavus also occurs from a remote (from assemblage 208/2) assemblage 275/1. In this case, belonging to a separate species is guaranteed with the very different coloration.

Derivation of name: *krajka* is Slavic for lace. Gender feminine.

Character of preservation: 1 clavus.

Assemblage 208/3

This assemblage contains indigenous rare *Ano nym* sp.n. (3791/94±) and common *Truhla vekov* described above.

Unidentifiable: 3791/103, 100 (forewings); 3791/103 (hindwing); 3791/97, 98, 93±. SUM = 6. 3791/96 does not belong to Blattaria or has been damaged during the deposition.

Assemblage 208/4

Unidentified cockroaches are: 3791/108 ± (complete specimen); 106 ± , 104F ± (2 immatures).

This assemblage entirely differs taphonomically, with occurrence of two unidentified immature specimens and also a complete adult. Most probably, this assemblage is without transportation and thus without decomposition and disarticulation of specimens. Otherwise, this assemblate is similar to previous, but not with others 208/ as it contains (a single) specimen of *Nuurcala cela* sp.n. described above. Nevertheless, a closely related separate sister species cannot be excluded based on this fragment.

Assemblage 203/6

This assemblage contains only a specimen of *Blattula flamma* sp.n. described above.

Assemblage 328

Unidentified cockroaches are: 3791/1205, 1202.

Blattoidea Latreille, 1810
Mesoblattinidae Handlirsch, 1906

Praeblattella Vršanský, 2003

Type species: *Piniblattella ponomarenkoi* Vršanský, 2003 (Böön Tsagaan Nuur).
Composition: *Praeblattella elegans* Vršanský, 1997 (Baissa), *Praeblattella dichotoma* Vršanský, 2003, *Praeblattella zrnko* Vršanský, 2003 (Böön Tsagaan Nuur); unpublished (New Jersey amber).
Stratigraphic range: Middle or Late Jurassic (present discovery)—Late Cretaceous.
Paleogeographic range: Laurasia.
Diagnosis (after Vršanský 2003): Small to medium size cockroaches. Head probably hypognathous (even often preserved in upright condition—similar to other Mesoblattininae), pronotum vaulted, often with complicated color pattern, slightly transversal, body depressed. Terminalia of females with very short external ovipositor (almost entirely internalised). Males with small tergal glands. Legs comparatively soft. Forewing margins parallel (exception is minute *P. zrnko* with 6.5 mm long wing); SC with 2–4 branched, RS may be differentiated. R is straight; M comparitevely rich with more than 4–5 branches; Cu branched, with three veins or more. Hindwing with R differentiated into R1 and RS, M usually simple or with 2–3 branches; CuA rich with several blind branches, CuP simple, possibly dichotomised near apex.

Praeblattella jurassica sp.n. (Fig. 6e).

Holotype: 3791/1026. A forewing fragment.
Differential diagnosis: It differs from congeners except *P. dichotoma*, its siter species (which differs only in having more extensively branched basalmost R) in having no coloured intercalaries (unlike *P. ponomarenkoi* and *P. elegans*) and in having no extremely simplified forewings (dislike minor *P. zrnko*).
Description: Except for the basis, forewing margins parallel, length ca. 13 mm. Costal area narrow, SC short, with three veins at margin. R1 with simply dichotomised branches and tertiary branched anteriormost branch, totally with 13 veins, RS differentiated with four or more veins at margin. M more or less straight, with about five veins at margin; cubital area long, with about five veins at margin.
Derivation of name: after Jurassic period. Gender feminine.
Character of preservation: one forewing.

Corydioidea Saussure, 1864
Chresmodidae Haase, 1890

Differential diagnosis: water-skimming cockroaches with multigemented tarsi in adult stage.
Type species: *Chresmoda obscura* Germar, 1839.
Composition: *Chresmoda* Germar 1839, *Jurachresmoda* Zhang et al. 2008, *Sinochresmoda* Zhang et al. 2008.
Stratigraphic range: Callovian—Cenomanian.
Geographic range: Cosmopolitan.

Remarks: Systematically unevaluated specimen 3791/1550 occur in unknown Bakhar layer. Chresmodidae are categorised on the basis of occurrence of skimming cockroaches immatures and *Chresmoda chikuni* Zhang et Ge 2017 adults in the Myanmar amber. Immatures do not possessed multisegmentation of tarsi, and thus, categorisation within this species and genus is not clear. Adult is insufficiently known, multisegmented cerci indicate it belongs to Blattaria, although multisegmented cerci although differently structured also occur in Eoblattida, Cnemidolestida and Reculida. Chresmodidae were already placed in Blattaria by Vršanský et al. (2019b).

Blattulidae Vishniakova, 1982
Blattula **Handlirsch, 1906**
Blattula bacharensis **sp.n**. (Figs. 1d, 15o).

Holotype: 3791/1203. A forewing fragment.
Paratype: 3791/1204. A forewing. Assemblage 328.
Type horizon: Assemblage 328.
Differential diagnosis. It differs from all representatives of the genus except sister taxon *B. choutinensis* in having partially colored membrane and wider veins. *B. choutinensis* membrane is partially transparent in all surface, is slightly larger in average (7 mm or more) and has distinct cross-veins.
Description: Forewing length ca. 7 mm or less. Wing margins parallel. Membrane partially colored, veins thick. Intercalaries present, cross-veins indistinct. SC simple and short, R nearly straight, with 13 veins meeting margin, RS undifferentiated. M with six veins at margin, slightly sigmoidal; CuA also slightly sigmoidal with six veins at margin.
Derivation of name: after Bachar (transliteration of Bakhar).
Character of preservation: two forewing fragments.

Context of Bakhar Cockroaches

General appearance of present taxa was highly conservative. All Blattulidae, Caloblattinidae, Liberiblattinidae and Mesoblattinidae are of standard appearance, with little morphological and size variation. Compared with various Cretaceous (and living) genera and families, no deviant, aberrant, bizarre or obscure taxa are present. The predatory *Raphidiomima* also occurs in diverse localities out of Bakhar. It cannot be excluded that laying isolated eggs and the female external ovipositor restricts eventualities of morphotypes. Nevertheless, disparity of forms is limited when compared with Paleozoic taxa (which includes Spiloblattinidae).

Sexual dimorphism of Mesozoic cockroaches is elaborated on the basis of Cretaceous taxa in Baissa and Böön Tsagaan (Vršanský 2003): some close species are known to possess counterindicative data. In some blattulid species, the male is bigger, in a closely related one, the female is bigger. So dimorphism is expected for the present taxa forming assemblages. Nevertheless, size difference and variability are superficial (Vršanský 2000) and present in both sexes, so dimorphism cannot bias the diversity presented in this study.

Moreover, there is no overlap of related species in respective assemblages. The only exceptions are *Rhipidoblattina sisnekharab* and *R. konserva*. Nevertheless, coloration and size difference is 200%. This case cannot be excluded with confidence and is of low probability based on experience from other sites. Occasional single taxonomic errors do not influence the general picture (of extreme diversity of 275/1), what more is that they both occur in 268 packages and not in the most diverse (275) site.

Assemblage analysis (see Fig. 22). Cockroaches reveal different stratification in different layers, namely differentiation of layers 203 (6); 208 (2, 3, 4); 268 (4, 8, 12,

The original version of this chapter was revised: Table 3 data replaced as dates, now it has been changed to number values. The correction to this chapter is available at https://doi.org/10.1007/978-3-030-59407-7_6

P. Vršanský, *Cockroaches from Jurassic sediments of the Bakhar Formation in Mongolia*, SpringerBriefs in Animal Sciences, https://doi.org/10.1007/978-3-030-59407-7_5

14, 19); 275 (1–2) and 328 without proved overlap (268 and 275 are strictly differentiated by different *Rhipidoblattina* species). Assemblage 203/6 reveals a single
specimen of *Blattula flamma*, which is normally found in 208 (2, 3), but due to a
high degree of similarity of this common genus, the transition cannot be proved. 208/4
reveals a single specimen of *Nuurcala cela* while common species of 208 are absent.
Species partition differs significantly among widely collected 208/2 and 208/3 so
that it is clear these assemblages are closely associated with three shared species, but
they represent different populations in different spatiotemporal conditions. Among
268 assemblages of numbers 4, 8, 12, 14, 19, there are no statistical differences and
they apparently represent closely associated beds and ecosystems. 275/1 and 275/2
represent another related bulk, differing to more extent than previous assemblages.
275/2 contains a high proportion of a single species, *Solemnia togokhudukhensis*,
caused either taphonomically or in differing real actuocenoses. These layers also
contain advanced *Perlucipecta,* suggesting a younger age. 328 contains a single
highly derived and advanced taxon found in Barremian of Mongolia, which is in
concordance with its younger age, comparable to Tithonian Khoutiin-Khotgor in
Mongolia. It is important that *Solemnia* from 275 assemblage is older than 328,
which is considered coeval with Khoutiin-Khotgor (but it contains not the same,
although very closely related *Blattula bacharensis*—a sister species with *Blattula
choutinensis*). Thus, other assemblages resemble the Tithonian. Otherwise, the most
advanced taxa *Raphidiomima* and *Ano* occur in the basalmost 208 as well as in 275
suggesting that the temporal difference was not significant and probably failed within
stages.

Taphonomy 275/1 is special regarding diversity as with 16 species it belongs,
globally, to the most diverse fossil assemblages (along with Myanmar amber and
Karabastau sediments). High diversity might be taphonomically caused and the single
bed may represent extensive time period with mixed actual populations; nevertheless,
it is intuitive as there are no overlaps of sibling species, suggesting that the fossilised
populations are close to the actual biota. It is remarkable that the dominant species
Ano da (based on forewings) lacks any isolated pronota (specimen 599 contains
forewing, both hindwings and a pronotum allowing categorisation of a pronotum
to this species) in the respective beds, while the bed contains dozens of isolated
pronota belonging to at least three other, unidentified species, which are not very
closely related. The assemblage 275/2 shares species *Ano da (n* = 3) and *Blattula
mini (n* = 1) with 275/1, but with different species ratios; under identical taphonomy,
this likely reflects climatic differences (see discussion). They also share numerous
Solemnia togokhudukhensis (n = 11) and an identical pronotum belonging to an
unknown species (1188).

Phylogenetic evaluation of respective species reveals no one primitive or
advanced species within the genera. Also, no advanced or primitive families (no
FOD, no LOD) were recorded. The site revealed three indigenous new genera (*Hra,
Polliciblattula* and *Okras*) and three genera for the first time (FOD: *Ano* J2-J3,
Perlucipecta J2-K2 and *Praeblattella* J2-K1). There is no data on the final occurrence
(LOD) among the rest of the genera (*Nuurcala* J2-K2; *Rhipidoblattina* J1-K2; *Ano*
and *Blattula* J1-K1; *Raphidiomima* J2-J3; *Caloblattina* J1-K1; *Liadoblattina* J1-K2;

Solemnia J2-J3). Thus, the evolutionary stage of the assemblages is found adequate with the proposed mid-Jurassic age for the assemblages 203–275 and unequivocally Upper Jurassic for 328 (this does not mean the age is Upper Jurassic, just that the cockroaches unequivocally indicate that this stage is closely related with Khoutiin-Khotgor in Mongolia). According to the presence of FOD, the age (of even 203–275) is rather early Late Jurassic rather than late Middle Jurassic. This is supported with a higher degree of similarity of even later beds in Upper Jurassic Karatau and Lower Cretaceous of Shar-Teg in Mongolia (*Blattula flamma* sp.n. specimen 3791/109 differs from *Blattula mongolica* Vršanský, 2004 only in its smaller size).

Within locality among bed plylogenetical analysis reveals consistency among assemblages either through whole profile (*Blattula* 203–328); through 208–275 (*Polliciblattula;* but *Ano, Raphidiomima* and *Nuurcala* are missing in 268) and through 268–275 (*Hra, Rhipidoblattina*) or reduced in *Truhla* (208), *Praeblattella* (328); *Perlucipecta, Solemnia, Okras, Caloblattina* (275); *Liadoblattina, Dostavba* (268). To summarise, through profile *Blattula* is not age-indicative (J1-K2). Assemblage 203 contains a single specimen *Blattula,* and thus this layer also does not bear any significant phylogenetic signal in this respect. On the other side of the column is 328 with the same genus, but also with *Praeblattella,* which is a significant advanced Mesozoic indicator, known only from the Cretaceous and thus this layer (bed, assemblage) differs from the rest of the profile. The other indigenous *Truhla* (208) within Blattulidae is a rather primitive taxon within Blattulidae with Early Jurassic and even with some Triassic affinities, so it represents evidence of the Middle Jurassic rather than the Upper Jurassic strata. *Liadoblattina* and *Dostavba* are limited to central assemblage 268 as they are widespread (T3-K2) or indigenous, it is not indicative. Taxa limited to 275 are equivocal (J1-K2; indigenous; J3; J3), but evidence of Upper Jurassic rather than Middle Jurassic age. Taxa shared from 275 and 268 are indigenous and non-indicative (T3-K2). Interestingly that taxa shared in both sides of the profile are missing in 268 and are rare (J1-K2, J3, indigenous), except common *Ano* (J3). Taking all this evidence together, and based on common species, cockroaches indicate the age of the Bakhar assemblages as Upper Jurassic. The Middle Jurassic age would mean the Bakhar cockroaches were extremely advanced at that time.

Parsimony analysis (all analyses are based on data in Table 2; Fig. 16). In the ten replicates, bootstrap topology identical to 1,000 replicates occurred, with slightly different clade supports. *Ano da* + *Ano da2* (two specimen of the same species) reveal 100% support (96.681 for 1,000 replicates); *cosmopolitana* + *jurassica* 75.79 (84.273 for 1,000 replicates); and 74.78 (81.752 for 1,000 replicates) support was for Caloblattinidae (62.49 for *sarko* + *togokhudukhensis*; 59.621 for 1,000 replicates); 50.13 for all others except *cela* (unsupported in 1,000 replicates). Other clades remained in polytomies. In 100 replicates, bootstrap also recognises clade *cosmopolitana* + *jurassica* (84.043) and Caloblattinidae (86.936; 57.273 for *sarko* + *togokhudukhensis*), but also with low probabilities (51.193; 50.322) recognises clades including all taxa except *nym, da, net, disko* (Liberiblattinidae) and except *nym, da, net, disko* and also *bavi*, respectively. Other clades also remained in polytomies. Generally, we can see that all the parsimony analyses recognise

Table 2 Character list of Bakhar cockroaches.

sp.	Bd	1	2	3	4	5	6	7	8	9	10	11	12	13	14	15	16	17	18	19	20	21	22	23	24	25	26	27
nym	208	0	?	0	1	0	?	?	0	0	1	1	?	0	1	1	1	1	1	1	?	1	1	1	1	1	0	0
da	275	0	0	0	1	0	1	0	1	0	1	1	1	0	1	1	1	1	1	1	1	1	1	1	1	1	1	1
da2	275	0	0	0	1	0	1	0	1	0	1	1	1	0	1	1	1	1	1	1	1	1	1	1	1	1	1	1
net	275	1	?	0	1	0	0	0	1	0	1	1	0	0	1	1	1	1	1	0	1	1	1	1	1	1	0	1
vulgara	275	0	0	0	1	0	1	0	0	0	1	1	1	0	1	1	0	1	0	1	1	1	0	0	0	0	0	1
mini	275	1	0	0	1	0	0	0	0	1	1	1	1	0	1	1	0	1	0	1	1	1	0	0	0	0	0	1
micro	275	0	0	0	1	1	1	0	1	1	1	1	1	0	1	1	0	1	0	1	0	1	0	0	0	0	0	1
flamma	203–208	1	0	0	1	0	1	0	0	1	1	1	1	0	1	1	0	1	0	1	0	1	1	1	1	1	0	1
universala	208	1	0	0	1	0	0	0	0	0	1	1	1	0	1	1	0	1	0	1	1	1	0	0	0	0	0	1
anuniversala	275	1	0	0	1	0	0	0	0	0	1	1	1	0	1	1	0	1	0	1	1	1	0	0	0	0	0	1
bacharensis	238	1	0	0	1	0	1	0	1	1	1	1	?	0	1	1	0	1	0	1	1	1	1	0	0	0	0	0
vremeni	275	0	0	0	0	0	0	0	0	0	1	0	0	0	0	1	0	1	1	0	0	1	1	1	1	0	0	0
pre	268	1	0	0	1	0	1	0	0	0	1	0	?	0	1	1	1	1	1	1	?	1	0	0	0	0	0	1
disko	275	0	0	0	1	0	?	0	1	0	1	1	?	0	1	1	1	1	1	0	?	1	1	0	1	0	1	1
bavi	275	0	1	1	1	0	1	0	?	0	1	1	?	0	1	1	1	1	1	0	?	1	0	0	0	0	0	0
nie	268	1	1	0	1	0	1	0	1	1	1	1	1	0	1	1	1	1	1	1	1	1	0	0	0	0	0	0
cela	275–208-268	0	0	0	0	0	0	0	0	0	1	0	0	0	0	1	1	1	0	0	0	1	1	1	0	1	0	1
sarko	275	1	1	0	0	0	?	0	?	0	1	0	0	0	0	1	0	0	0	0	0	1	1	0	1	0	0	0
cosmopolitana	275	0	1	0	1	1	1	0	1	0	1	0	1	0	1	1	0	1	0	1	1	0	0	1	0	0	1	0
analis	208	1	0	0	1	1	1	0	1	1	1	1	1	1	1	1	0	1	1	1	1	0	0	0	0	0	0	1
tatosanerata	268	1	0	0	1	0	1	0	1	1	0	0	?	1	1	1	1	1	0	1	?	0	0	0	0	0	0	1

(continued)

Table 2 (continued)

sp.	Bd	1	2	3	4	5	6	7	8	9	10	11	12	13	14	15	16	17	18	19	20	21	22	23	24	25	26	27
jurassica	238	0	1	0	1	0	1	0	1	0	1	0	1	0	1	1	0	0	0	0	1	1	1	1	0	0	0	0
chimnata	208	0	0	1	1	0	1	1	0	0	1	1	1	0	1	1	1	1	0	1	1	1	0	0	0	0	0	1
bakharensis	275	0	0	0	0	0	0	0	0	0	1	0	0	0	0	1	0	1	0	0	0	1	1	0	0	0	0	0
sisnerahkab	268	1	0	0	1	0	0	0	0	0	1	0	0	0	0	1	0	1	0	0	0	1	1	1	0	0	0	0
konserva	268	0	0	0	0	0	0	0	0	0	1	0	0	0	0	1	0	1	0	0	0	1	1	1	0	0	0	0
togokhudukhensis	275	1	11	0	1	0	0	0	0	0	1	0	0	0	0	1	1	1	0	1	0	1	1	1	0	0	0	0
vekov	208	0	0	1	1	0	0	1	0	0	1	1	?	0	1	1	1	1	0	1	?	0	0	0	0	0	0	1

sp.	Bd	28	29	30	31	32	33	34	35	36	37	38	39	40	41	42	43	44	45	46	47	48	49	50	51	52	53	54
nym	208	0	1	0	1	0	1	0	1	1	1	1	0	0	0	0	0	?	?	1	0	0	0	?	?	0	?	?
da	275	0	1	0	1	0	1	0	1	1	1	0	0	1	0	0	0	1	1	1	0	0	1	0	1	0	0	0
da2	275	0	1	0	1	0	1	0	1	1	1	1	0	1	0	0	0	1	1	1	0	0	1	0	1	0	0	0
net	275	0	1	0	1	0	1	0	1	1	1	1	0	0	0	0	0	1	0	1	0	1	0	0	0	0	0	1
vulgara	275	0	0	0	0	0	1	0	0	0	1	0	0	1	0	0	1	0	0	1	0	0	0	?	1	0	?	?
mini	275	0	0	1	0	1	1	0	1	0	1	0	0	0	1	1	1	0	0	1	0	0	0	1	1	0	1	1
micro	275	1	0	1	0	1	1	0	1	0	1	0	0	1	0	0	1	0	0	1	0	0	0	1	1	0	1	1
flamma	203–208	0	0	1	1	1	1	0	1	1	1	1	0	1	0	0	0	0	0	1	0	1	0	1	1	0	1	1
universala	208	0	0	1	0	1	1	0	1	1	1	1	0	1	0	0	0	0	0	0	0	0	0	1	1	0	1	1
anuniversala	275	0	0	1	0	1	1	0	1	1	1	0	0	1	0	0	0	0	0	1	0	0	0	1	1	0	1	1
bacharensis	238	1	0	1	0	1	1	0	1	1	1	0	0	1	0	0	0	1	0	1	0	0	0	1	1	0	?	1
vremeni	275	0	0	0	0	0	0	0	0	0	0	0	0	0	1	0	0	0	0	0	0	1	1	0	0	0	0	0
pre	268	0	1	0	0	0	0	1	1	1	1	1	0	1	0	0	0	0	0	1	0	0	0	0	0	0	0	0

(continued)

Table 2 (continued)

sp.	Bd	28	29	30	31	32	33	34	35	36	37	38	39	40	41	42	43	44	45	46	47	48	49	50	51	52	53	54
disko	275	0	1	0	1	?	1	0	1	1	1	1	0	0	0	0	0	?	?	1	0	1	0	?	?	?	1	?
bavi	275	0	1	0	1	0	1	0	1	0	1	1	0	0	0	?	0	?	?	1	0	1	0	?	?	?	1	?
nie	268	1	0	0	0	1	1	0	1	1	1	1	0	1	0	0	0	0	0	1	0	1	0	1	1	0	1	1
cela	275–208–268	0	0	0	0	0	1	0	0	0	1	0	0	0	1	1	0	1	0	0	0	0	0	0	0	0	0	0
sarko	275	0	0	0	0	1	0	0	0	0	1	0	0	0	0	0	0	1	0	0	0	0	0	0	0	0	0	1
cosmopolitana	275	0	0	0	0	1	0	0	1	1	1	0	0	0	0	0	0	0	0	0	0	1	0	1	0	0	1	1
analis	208	1	0	1	1	0	0	0	1	1	1	1	0	1	0	0	1	0	0	1	0	0	0	1	1	0	0	1
tatosanerata	268	0	0	1	0	0	0	0	1	0	1	0	0	1	0	0	0	0	0	1	0	0	0	1	?	?	?	?
jurassica	238	0	0	0	1	0	0	0	1	0	0	0	0	1	0	0	0	0	0	0	0	2	0	2	0	0	0	1
chimnata	208	1	0	0	1	0	1	0	0	0	1	0	0	0	1	1	1	1	0	1	1	1	0	0	0	0	0	0
bakharensis	275	0	0	0	0	0	0	0	0	0	1	0	0	0	0	0	0	1	0	0	0	0	0	0	0	0	0	0
sisnerahkab	268	0	0	0	0	0	0	0	0	0	1	0	0	0	0	0	0	1	0	0	0	1	1	0	0	0	0	0
konserva	268	0	0	0	0	0	0	0	0	0	0	0	0	0	0	0	0	1	0	0	0	0	0	0	0	0	0	0
togokhudukhensis	275	0	0	1	1	1	0	0	0	0	1	0	0	1	0	0	0	1	1	0	0	0	0	0	0	0	0	1
vekov	208	0	0	1	0	0	0	0	1	1	1	1	1	1	0	0	0	0	0	1	0	?	0	?	?	0	?	?

Caloblattinidae. Some analyses (100 replicates) also weakly support most Liberiblattinidae, while Blattulidae remain unrecognised among Raphidiomimidae and
Mesoblattinidae. Reliable 1,000 replicate bootstraps recognise only (most basal)
Caloblattinidae, while all others remain in polytomy. These analyses, in contrast to
intuitive analyses of genera and families, do not reveal any signal related to distribution in bed, respectively, in particular assemblages or their packages. Nevertheless, a significant pattern validated for main cockroach lineages (Sendi et al. 2020a,
b, Vršanský et al. 2018, 2019), namely the "explosively radiating" reduction ring,
is observed also in the present data within respective families and evidence rapid
appearance of genera during the family-origin time.

Bayesian network analysis is far more informative (Fig. 18). All species except
liberiblattinid *H. nie* (housed with insignificant support inside separated Blattulidae) are recognised as housed within their families (Caloblattindae 87; Liberiblattinidae 77.1; Mesoblattinidae 95). Raphidiomimidae with a single species has 100%
support. Within Caloblattinidae, supported are *sarko + togokhudukhensis* (63.6); all
except *togokhudukhensis* (55.9) and all except *sarko + togokhudukhensis* (63.7);
bakharensis + conserva (68.9). Within Liberiblattinidae supported is *Ano da + Ano
da2* (99) and *Ano da, Ano da2, Ano nym + Ano net* (66.3).

Regarding the respective packages of assemblages, uppermost 328 is distinct
possessing one of two advanced Mesoblattinidae. Lowermost 203–208 is distinct
for possessing a single Raphidiomimidae; nevertheless, this family occurs also in
268 and 275, but these are represented by fragments which were not included in the
analysed matrix. Distribution of other taxa is not informative in respect of different
beds within Bakhar.

From the more general perspective, corydioid lineage (sensu Vršanský 2002) is
supported, with families Blattulidae and Liberiblattinidae largely (but still insignificantly) represented. Also insignificantly represented are Liberiblattinidae, with
Caloblattinidae, which reflects their shared ancestry via *Volziablatta* group or
Phyloblattidae. Raphidiomimidae, which belongs to Caloblattinoidea (Vršanský and
Bechly 2015) is insignificantly represented with Caloblattinidae. Mesoblattinidae are
significantly represented with Caloblattinidae which reflects their origin (Šmídová
2019, 2020). Inconsistent (but insignificant) approximation of Raphidiomimidae with
Blattulidae is related to similar morphology of clavus. It must be considered that these
results are based on forewings only while the hindwing is phylogenetically even (far)
more informative.

Network analyses within respective beds (Fig. 17) were performed to access
data that are usually obtained but are often not accessible. They reveal taphonomic
bias of respective isolated beds. Nevertheless, similarly represented assemblages 275
and 268 ($n = 540, 575$) at first sight appear to be most diverse in 275.

Assemblage 268 reveals highly consistent Caloblattinidae (100), 97.5 with two
species, *Rhipidoblattina sisnekharab* and *Rhipidoblattina konserva; konserva* with
cela (94). 92 represent Liberiblattinidae (*tatosanerata* and *Hra nie*).

Within assemblages 203 + 208, supported are Blattulidae (*Polliciblattula analis,
Truhla vekov, B. flamma, B. universala*; 94.8) and also a clade consisting of *Nuurcala cela* (Caloblattinidae) + *Ano nym* (Liberiblattinidae) (96.8). At the same time,

Nuurcala cela with *Raphidiomimima chimnata* (Raphidiomimidae) is distinguished from other taxa (94.1). Within Blattulidae, *Blattula flamma, B. universala + Polliciblattula analis* are separated from *Truhla vekov* (89.6); *flamma + universala* (69) and *analis + flamma* (92.2).

The more extensive 275 strongly support intuitively discriminated families (Caloblattinidae, 93.1; Blattulidae, 96.4; Liberiblattinidae 71.9). Two specimens of *Ano da* have 99.7 support (differ in one character). Genus *Ano* with *Hra disko* have 80.7 support. Within Blattulidae, *B. vulgara* is discriminated (78), and *B. micro* and *B. vulgara* are also supported (52.7). Within Caloblattinidae *Rhipidoblattina bakharensis, Caloblattina vremeni* and *Nuurcala cela* (84.5); *Okras sarko* and *Solemnia togokhudukhensis* (56.4); *Solemnia togokhudukhensis* is distinct from the rest Caloblattinidae (57.9). *Perlucipecta cosmopolitana* (Mesoblattinidae) is a typical long branch in this restricted analysis.

Age separation within locality, among uppermost and lowermost beds as mentioned above, was considered. The bottommost layer 203 contains only one specimen of the genus *Blattula*, so it is not representative. Uppermost layer 328 contains the Cretaceous genus *Praeblattella*, which is the only Cretaceous indicator in the entire assemblage, so it might appear rather significant regarding the age difference from the rest of the assemblages. Additionally, the species is closely related to an Aptian representative of this genus from Böön Tsagaan in Mongolia. Representatives of the genus *Blattula* reveal the time difference, if their presence is not considered as *Blattula* is extremely rare in the Cretaceous and as a matter of fact is present only in the Chernovskie Kopi in Russia (Barna 2014). Rough age identity is postulated for all insects, especially the large ones such as *Karabasia caudata, Bakharia gibbera, Cyclothemis sp.* and *Perlariopsis fidelis* which are together, while differences include the presence of mayflies, crickets and *Taeniopteryx,* and the absence of caddis cases and rakes (Ponomarenko 2019). Age separation of all other beds is apparently insignificant. This is documented not only by the presence of shared genera, but most importantly by representatives of *Polliciblattula* indigenous to the locality and found in different beds. Differing species suggest speciation can slightly span over a million years, but with a species not significantly lasting more than ten. Differences of *B. universala* and *B. anuniversala* (203/275) are minor. Age differences among beds in respective packages (208; 268; 275) are superficial, as documented with most shared species. Details are provided in the taphonomical paragraph.

A closely related site that is possibly identical with assemblage 328 is Khoutiin-Khotgor adjacent in Mongolia. In addition to sharing a sister taxa within these two localities, *Solemnia* (only from assemblage 275!) and *Blattula*, in Khoutiin-Khotgor, as well as the indigenous genus *Irreblatta* Vršanský, 2008 that is categorised within "*Voltziablatta*-group", commonly occur. This family is restricted to Khoutiin-Khotgor during the Jurassic, otherwise (commonly known as a cosmopolitan lineage) occurs only in the Triassic. So while Khoutiin-Khotgor is roughly coeval and contains closely related taxa, it is also more unique compared with Bakhar 328 and in general.

Adjacent Chinese Jurassic sites contain mostly Caloblattinidae and Blattulidae, namely *Samaroblattula*; *Sogdoblatta* (Yujiagou); *Taublatta* (Yangshugou) and *Euryblattula* (Kuntouyingzi and Mayingzi). *Taublatta* and *Euryblattula* are present also in Guanyintan (Lin 1986). Diverse *Taublatta* are present in Shiniuwei (Lin 1986). Obscure *Jingyuanoblatta* and *Samaroblatta* are recorded in Wangjiashan (Lin 1982). Extensively collected cockroaches at sites in Xiaofanzhangzi (Hong 1980, 1983, Zhang 1986), Meitian (Lin 1986), Zhouyingzi (Hong 1980; Ren et al. 1985; Zhang 1986) and Huapen (Hong 1997; Hong and Xiao 1997) provide some clues. At Xiaofanzhangzi is found *Blattula, Euryblattula, Mesoblattula, Rhipidoblattina* and *Rhipidoblatta* (designated originally as *Triassoblatta*). Meitian contains *Caloblattina, Rhipidoblatta, Perlucipecta, Blattula, Rhipidoblattina, Hichuja* and *Nuurcala*. Zhouyingzi contains *Parablattula, Rhipidoblattina, Sogdoblatta, Fusiblatta, Perlucipecta,* and *Samaroblatta*. Huapen consists of *Samaroblatta, Sogdoblatta, Rhipidoblattina* and *Euryblattula*. We see that all these sites are similarly structured with a dominance of Caloblattinidae, Blattulidae and Liberiblattinidae (but coloured liberiblattinids that dominate in Bakhar were absent). Mesoblattinidae, Chresmodidae and Raphidiomimidae were absent or extremely rare, also in the adjacent sites.

Comparing the Middle Jurassic assemblages, Russian site Kubekovo contains only *Blattula, Chiloblattula, Eublattula* (Blattulidae; Vishniakova 1982, 1985), making Bakhar assemblages phenetically dissimilar as lower beds contain Liberiblattinidae and upper beds commonly contain Caloblattinidae. Although we did not completely evaluate Daohugou, likely evidence shows a comparative partition of Blattulidae, while advanced sophistically coloured Liberiblattinidae were very rare in Daohugou (Vršanský et al. 2012). On the other hand, we found a huge diversity of predatory Raphidiomimidae (Liang et al. 2009a, b, 2012) in Daohugou. Raphidiomimidae are more advanced in Bakhar, while advanced Raphidiomimidae were absent in Daohugou. Therefore, it seems Bakhar is younger in age compared with Daohugou. Fuziidae were indigenous to Daohugou, although widespread and with more than 12 (7 formalised) species of diverse size and coloration. Other Middle Jurassic sites are not informative but contain *Blattula* (Lin 1986); *Elisama* (Wang 1987); *Fusiblatta* (Hong 1980); *Mesoblatttina* (Hong 1980; Lin 1982, 1986; 1985; Ping 1928); *Mesoblattula* (Zhang 1986; Hong 1986); *Rhipidoblattina* (Hong 1980; Hong 1983; Lin 1986; Zhang 1986; Lin 1987); *Samaroblatta* (Vishniakova 1985); *Samaroblattula* (Hong 1982; Vishniakova 1985; Hong and Xiao 1997); *Sogdoblatta* (Hong 1983; Vishniakova 1983) or *Taublatta* (Hong 1983; Vishniakova 1985).

Jurassic context: The earliest significant Jurassic assemblages are from Toarcian in Germany and England, as summarised by Vršanský and Ansorge (2007). These sites are similar in diversity with Blattulidae, specifically *Blattula dubia* (Handlirsch 1939) resembling *Polliciblattula* in size and morphology. These two sites are similar for Caloblattinidae (*Caloblattina, Rhipidoblattina, Liadoblattina*, while *Nuurcala* is characteristic for later assemblages and occurs predominantly in the Cretaceous). Significant differences are abundance with advanced Mesoblattinidae, which are extremely rare in Bakhar, but present as different genera (*Mesoblattina* Geinitz, 1880) in Toarcian. Comparable liberiblattinid species to *Ano* is *Eublattula* Handlirsch, 1939 from the Toarcian.

Similar assemblages are the Early Jurassic Mintaja of Australia (Martin 2010), with shared *Blattula, Elisamoides, Caloblattina, Falcatussiblatta* Liang et al. 2018, and *Rhipidoblattina*. Liberiblattinid *Kurablattina* Martin, 2010, is a unique and advanced taxon, similar as found in Toarcian sites, while Mesoblattinidae were missing in Mintaja.

Karabastau is only partially represented, with shared *Blattula, Liadoblattina, Nuurcala, Perlucipecta, Praeblattella, Rhipidoblattina, Truhla* and *Raphidiomima* (Vishniakova 1968, 1971). In Karabastau, *Solemnia, Dostavba* and *Hra* are missing, while a high number of other genera are present. Represented were *Srdiecko* Vršanský 2008; *Paleovia* Vršanský 2008; *Skok* Vršanský 2007; *Decomposita* Vršanský 2008; *Artitocoblatta* Handlirsch 1906; *Asioblatta* Vishniakova 1968; *Rhipidoblattinopsis* Vishniakova 1968; *Latiblatta* Vishniakova 1968; *Karatavoblatta* Vishniakova 1968, while many others need description (Vishniakova 1968, 1971, 1973; Vršanský 2007, 2008).

Shurab (see Martynov 1937) consists of *Caloblattina, Solemnia, Taublatta, Raphidiomima, Blattula, Rhipidoblattina, Perlucipecta, Samaroblattula* and *Mesoblattula,* with at least three distinctly different genera.

Regional Jurassic sites in China (other than Middle Jurassic mentioned in the paragraph above) include Haifanggou, Lamagou, Chicheng, Hongliu Geda, Pengzhuang, Dakouzi, Kuntouyingzi, Dameigou, Dongchangtai are fragmented which hamper a deep comparison, but they are not contradictory in the composition and/or do not possess indigenous taxa and/or taxa different from Bakhar (Hong 1980, 1982, 1983; Wang 1987; Lin 1978, 1985; Hong and Xiao 1997). Notably, *Perlucipecta* is present in Pengzhuang.

Middle Jurassic Kubekovo in Siberia contains *Blattula aberrans* Vishniakova, 1982 (complete specimen) and *Chiloblattula sibirica* Vishniakova, 1985 (hindwing). Both sites contain the genus *Blattula.*

To summarise, Bakhar represents a more or less standard Jurassic assemblage with the presence of species of Jurassic stage and also several indigenous taxa.

Indigenous genera are limited at Bakhar. There are no endemic genera present in Bakhar. Some of the herein described genera occur in at least one undescribed Jurassic fauna. The common occurrence of indigenous taxa in the Jurassic is remarkable. Indigenous *Blattulites* and *Ijablatta* in Iya (Vishniakova 1982, 1983); *Etapia* in Russian Chernyy Etap II (Vishniakova 1983); *Irreblatta* and *Solemnia* in Mongolian Khoutiin-Khotgor (Vršanský 2008); *Itchetuja* in Russian Novospasskoe (Vishniakova 1983); *Kemerowia* in Russian Lyagush'ye (Vishniakova 1983); *Schambeloblattina* in Schambelen, Switzerland (Handlirsch 1906–1908); *Soliblatta* and *Summatiblatta* in Chinese Meitian (Lin 1986); *Taublattopsis* in Mongolian Dzhargalant (Vishniakova 1985); *Kurablattina* in Mintaja (Martin 2010); *Jingyuanoblatta* in Chinese Wangjiashan (Lin 1982). (Note that Mongolian Shar-Teg is considered for Cretaceous (Vršanský et al. 2017).)

Speciation is especially expressed in the well-studied Daohugou (Liang et al. 2009, 2012, 2019; Vršanský et al. 2009, 2012; Wei et al. 2012, 2013; Guo and Ren 2011), where the entire, large indigenous family Fuziidae (*Arcofuzia, Colorifuzia, Fuzia, Parvifuzia*) occurs. Nevertheless, in Daohugou, there are numerous indigenous genera of the predatory family Raphidiomimidae (*Divocina, Fortiblatta, Gracilliblatta, Falcatusiblatta*). Speciation of Liberiblattinidae (*Entropia)* is not surprising as most of the genera within this family are indigenous. Numerous indigenous genera from Daohugou await description, including indigenous genera of the conservative Blattulidae.

Genera not indigenous are widely represented outside of Bakhar. *Mesoblattina* occurs in Toarcian of Germany, in Zhouyingzi and in Karabastau (and possibly in Swiss Schambelen according to Handlisrch 1906); *Lithoblatta* in Solnhofen and Karabastau; *Rhipidoblatta* in Karatau, Kuntouyingzi, Zhouyingzi, Dongchangtai, Daokuntouyingzi and numerous Cretaceous sites; *Taublatta* in Oshin-Boro-Udzyur-Ula, Guanyintan, Iya, Ninghua, Shiniuwei, Guanyintan and Yangshugou; *Sogdoblatta* in Voznesenye, Ust-Baley, Yujiagou, Huapen and Zhouyingzi; *Samaroblatta* in Lyagush'ye, Dongsheng, Wangjiashan, Huapen, Oshin-Boro-Udzyur-Ula, Meitian and Zhouyingzi (and also in Siberia in South Africa); *Parablattula* in Iya, Zhouyingzi, Uda. *Mesoblattula* occurs in Xiaofanzhangzi, Iya and Xiwan. *Euryblattula* is reported from Kuntouyingzi, Mayingzi, Huapen, Meitian, Xiaofanzhangzi and Guanyintan. *Samaroblattula* occurs from Huapen, Dameigou, Zaoshang and Oshin-Boro-Udzyur-Ula. *Eublatula* in Kubekovo and Dobbertin.

Paleogeographically, most taxa belong to cosmopolitan genera (five species in genus *Blattula, 1 Caloblattina, 1 Praeblattella, 1 Perlucipecta*), Laurasian *(1 Nuurcala, 1 Liadoblattina, 3 Rhipidoblattina)*, regional (*Solemnia:* this genus occurs in Mongolian Khoutiin-Khotgor, *2 Raphidiomima:* this genus occurs in Karabastau Formation, *3 Ano:* this genus occurs in Karabastau Formation) or indigenous (*3 Hra, Okras, Polliciblattula, Truhla*). A single circumtropic taxon is indicated with an isolated pronotum probably belonging to tropical Laurasian (in the widest sense) Myanmar amber and tropical Gondwanan Crato (Vršanský et al. 2019). Exclusively, Gondwanan taxa or taxa found only in Myanmar amber are also missing. Due to well-studied global context of Jurassic-Cretaceous cockroaches, this pattern is standard except for the occurrence of regional taxa. If we disregard the occurrence of *Solemnia* known from related Khoutiin-Khotgor strata, two genera shared with Karabastau formation found nowhere else rather indicate a coeval sedimentation rather than a close geographic position. Karabastau in Kazakhstan and Bakhar in Mongolia lies closely together, both with a warm climate; however, this similarity barely reflects topographic approximity. Due to the rapid and active spread of cockroaches, the findings represent further support for the Upper Jurassic age of Bakhar, like Karatau. Moreover, Mongolia, except for its westernmost part, and Karabastau belonged to different biogeographic areas in the late Mesozoic (Ponomarenko and Popov 2016) and thus a similarity in composition more strongly suggests a similar age. This regional finding rather strongly indicates a similarity with the Tithonian Khoutiin-Khotgor, regarding strata 275 and below. Interestingly, the genera widely distributed (*Caloblattina, Praeblattella, Perlucipecta, Nuurcala, Liadoblattina, Rhipidoblattina*) later radiated or evolved to new genera in the Cretaceous. On the contrary, higher, family-rank radiation took place only from the rare and indigenous Liberiblatindae (Vršanský et al. 2017).

Botanical provinces, namely the northern West Siberian Province of the Siberian Region, which characterises the Bakhar flora, and the North Chinese Province South of Bakhar, do not seem to be identified by cockroaches, and there is little association.

On the contrary, cockroaches seem to follow provinces preferred by Karabastau Fm., namely Euro-Sinian paleofloristic province, which is characterised by the abundance of Bennettitales (significantly with "flowers"), Cheirolepidiaceae and the scarcity of Czekanowskiales and Ginkgoaceae according to Vakhrameev (1991).

Numerical comparisons using **phenetical analyses (Figs. 20–21; all locality analyses are based on occurrence of genera in Tables 2 and 3)** reveal the similarity of Bakhar and Karabastau using both the hierarchical clustering (40%) and neighbour joining methods with Jaccard indexes (Fig. 21), with a different degree of similarity of the Daohugou (see difference among Fig. 21a, b). Approximity of Early Jurassic Mintaja and Dobbertin with Cretaceous Chernovskie Kopi rather express advanced assemblages in Germany and Australia rather than primitive assemblage in Russia. Mintaja (Chernovskie Kopi 80%) especially contains an advanced taxa found mainly in the Upper Jurassic. On the other hand, cockroach assemblages in Chernovskie Kopi are exclusively Late Jurassic, although other groups such as termites provide evidence of an Early Cretaceous age. These two factors together result in their approximity in the present analysis. It might be helpful to consider a similar parallel in Xiangshan, where a putative angiosperm occurs in Early Jurassic sediments (Fu et al. 2018). Remarkable in these first analyses are (basal) approximations of Khoutiin-Khotgor and also Shar-Teg, which is Cretaceous, showing the temporal relationship of Bakhar and Karabastau close to the J3 rather than J2. Latter analyses approximate Daohugou and Meitian (distinctly J2).

In correspondence ordination analysis, Bakhar is topographically related to Meitian and Xiaofangshanzhi, while Karatau is located more closely to Cretaceous Chernovskie Kopi and Shar-Teg (Fig. 20; numbers in both a and b analyses represent the same assemblages explicitly listed in Fig. 20a). This more complex analysis is very difficult to interpret. While the triangle formed by Bakhar, Karabastau and Daohugou might explain their complex spatiotemporal relationship, the position of this triangle is equivocal. Bakhar seems to have shifted (logically) to Khoutiin-Khotgor as well as (more difficult to interpret) to Early Jurassic Mintaja and Dobbertin (and also above-mentioned Chernovskie Kopi) on one side and Shurab on another side. Khoutiin-Khotgor is more distant which results from very limited data at the analysed (genus) level. Shar-Teg is very remote, which likely corresponds to a characteristic, different, Cretaceous stage of this assemblage (Cretaceous Chernovskie Kopi is an assemblage of Upper Jurassic type). This analysis does not reflect unequivocally an age, nor a topographic position, but might be useful in wider future comparison above the present order level.

Performed for information purposes (phenetical similarity) were also **neighbour joining clustering using Jaccard indexes for respective species**. This analysis (Fig. 19) reveals a close approximity of all Caloblatttinidae and Mesoblattinidae, respectively. Separated (and supported) are also Liberiblattinidae and Blattulidae except for *Dostavba pre* and *Hranie* (Liberiblattinidae) housed inside Blattulidae.

A Triassic link to Bakhar assemblages is lacking. Triassic assemblages consistently contain Subioblatttidae, *Volziablatta* group apparently absent in Bakhar, but the latter recorded in Khoutiin-Khotgor which is its only one Jurassic locality. They also

Table 3 Phenetic similarity of major Jurassic sites with Cretaceous Chernovskie Kopi and Shar-Teg

	n	Bak	Kar	Mei	Zho	Hua	Xia	Dao	Dob	Min	Kho-Kho	Shu	Sha	Che
Shurab	37	6/9	6/9	4/9	5/9	3/9	3/9	6/9	4/9	3/9	2/9	–	1/9	2/9
Dobbertin	119	5/6	6/6	4/6	4/6	4/6	4/6	4/6	–	3/6	2/6	4/6	1/6	3/6
Mintaja	374	4/5	4/5	3/5	2/5	2/5	3/5	3/5	3/5	–	2/5	3/5	3/5	4/5
Daohugou	1500	8/16	3/16	5/16	3/16	4/16	3/16	–	4/16	3/16	2/16	6/16	1/16	3/16
Meitian	?	6/7	6/7	–	4/7	4/7	3/7	5/7	4/7	3/7	2/7	4/7	1/7	4/7
Zhouyingzi	?	4/7	5/7	4/7	–	3/7	2/7	3/7	4/7	2/7	2/7	5/7	2/7	2/7
Huapen	?	3/4	4/4	4/4	3/4	–	3/4	4/4	4/4	2/4	2/4	3/4	1/4	3/4
Xiaofanzhangzi	?	3/4	4/4	3/4	2/4	3/4	–	3/4	4/4	3/4	2/4	3/4	1/4	3/4
Bakhar	1179	–	11/16	6/16	4/16	3/16	3/16	8/16	5/16	4/16	2/16	6/16	2/16	4/16
Khoutiin-Khotgor	20	2/3	2/3	2/3	2/3	2/3	2/3	2/3	2/3	2/3	–	2/3	1/3	2/3
Karabastau	2175	11/77	–	6/77	5/77	4/77	4/77	3/77	6/77	4/77	2/77	6/77	4/77	4/77
Shar-Teg	170	2/6	4/6	1/6	2/6	1/6	1/6	1/6	1/6	3/6	1/6	1/6	–	6/6
Chernovskie Kopi	41	4/6	4/6	4/6	2/6	3/6	3/6	3/6	3/6	4/6	2/6	2/6	6/6	–

consistently contain a small-sized advanced representative of the family Phyloblattidae, which was also absent in Bakhar and in other Jurassic sites (although relictual this family might be preserved at Böön Tsagaan). The last Triassic group, Caloblattinidae was preserved in Bakhar, but cannot be considered for a Triassic relic due to consistent occurrence of the family throughout the whole Triassic, Jurassic and Cretaceous, disregarding the climatical zones.

There is no special link to Cretaceous cockroach genera in Bakhar. Among 243 registered Cretaceous cockroach species in EDNA (fossil insect database active 2019-11-03), shared genera are only widely spatiotemporary distributed *Blattula* (Purbeck, Gaositai, Wiltshire, Chongqing, Chernovskie Kopi), *Rhipidoblattina* (Purbeck, Chaomidian, Housghah, Gurvan-Ereniy-Nuru, Xiagou, Chernovskie Kopi, Baissa), *Caloblattina* (Baissa, Böön Tsagaan), *Nuurcala* (Baissa, Böön Tsagaan, Khurilt, Huangbanjigou) and *Perlucipecta* (Baissa, Crato, Yixian, Myanmar) (Scudder 1886; Lin 1978; Hong 1982; Ren 1985; Vishnyakova 1986; Vršanský 2003, 2008; Wang et al. 2007; Wei and Ren 2013; Barna 2014; Lee 2016). Well-evaluated Cretaceous sites [Purbeck, Wealden, Baissa, Böön Tsagaan, Yixian with Laiyang (Chen et al. 2019), Crato, Chernovskie Kopi, Montsec (Martínes-Delclós 1993)], in contrast to Bakhar, mostly contain newly (at 127 Ma diversification point) originated mantodeans, termites, alienopterids, umenocoleids, ectobiids, while *Volziablatta* group went extinct. Specialised Cretaceous taxa thus apparently originated much later, at J/K or directly at 127 Ma diversification point (Vršanský et al. 2017), and there is not a single advanced in sense of Cretaceous taxon known from Bakhar. Surprisingly, primitive *Archimesoblatta* known from most Cretaceous sites in both Laurasia and Gondwana (Crato, Myanmar, Baissa, Böön Tsagaan, Chernovskie Kopi) and the lowermost Jurassic of Connecticut (Huber et al. 2003) are absent in Bakhar.

Special Cretaceous amber (Myanmar, Lebanon, Taimyr, Archingeay, New Jersey) differed significantly and contains additionally Blattidae, Manipulatoridae, Olidae, Socialidae and Eadiidae. The only shared genus is *Perlucipecta* known from Myanmar, Lebanese and Taimyr ambers. Generally, in ambers, bark cohorts and small species are (co)dominantly represented. In Cretaceous, taxa with ootheca start to dominate (see Hinkelman 2019; Gao et al. 2019), including taxa that hold ootheca permanently. Notably, in Bakhar, taxa with ootheca already existed (*Perlucipecta, Praeblattella*), but did not become common. Likely, the parasite load to oothecaholders was low. Interestingly, during Cretaceous, taxa without protected eggs still persisted as dominant (Caloblattindae and Blattulidae), mostly represented with the same genera (but *Elisama* becomes dominant blattulid genus during Cretaceous). Liberiblattinidae retained their position as a rare but diverse group during both the Jurassic and Cretaceous and also in other ambers.

Climatic inferences are in concordance with the proposed (Ponomarenko and Popov 2016) climatic maximum. It is supported with high diversity but also with occurrences of new disparity forms such as the dominant *Ano*. The same concerns rich coloration with high proportion of dark parts supporting either high temperatures or high humidity or combined.

Regarding respective beds, 275/1 might be warmer than 275/2 as it contains a higher proportions of primitive *Blattula* and *Caloblattina*. Variability (see *Ano* descriptions) also reveals shared 275/1 and two species differing at population levels and does not contradict warmer 275/1 (more collected specimens might also suggest a more productive beds).

Other sets of assemblages differ insignificantly.

Coloration of several species was sophisticated, indicating that the sophisticated stage had an advanced ecology with a more opened lifestyle. It is interesting that the style of coloration of cockroaches differed during different times. Permian cockroaches already contained a variety of colour patterns including zebra-like Phyloblattidae and sophistically patterned (with sophisticated stripes) Spiloblattinidae (including Subioblattidae). Coloration is restricted to stripes, but is so typical which can be used in stratigraphy (Schneider and Werneburg 1993). The coloration of Triassic cockroaches occurring at the beginning of the Mesozoic was very simple or absent for a significant time period. Only in late Middle Triassic (Madygen) did striped cockroaches occur again. Coloration of Early Jurassic cockroaches was rare and also restricted to more or less simple stripes like in *Liadoblattina*. Middle Jurassic is the time when sophisticated coloration began to spread. Daohugou cockroaches contain a similar diversity of colorations as in Bakhar, although not as frequent or expressed. This can be caused by a more advanced evolutionary stage of Bakhar or its more humid and warmer climate in lower paleolatitutes. In Bakhar, new types of coloration are demonstrated by a huge diversity of pronotal colorations (Figs. 2, 3). We observed total dark colorations (at least black and brown), simple stripes, sophisticated stripes, sophisticated carved patterns, simple maculas and minor maculation. In wings, coloration along veins also occurs. Dotted patterns, aposematic or camouflaged like seeds appear. Late Jurassic cockroaches already contained all coloration patterns found in living ecosystems, except for luminescence and possibly for total transparency.

Wing areas were calculated in respect of Oružinský and Vršanský (2017) in order to access area trends in ten complete forewing specimens with completely preserved venation.

Ano da specimens 3791/877 and 474 reveal (5.8 mm long) area 9.1 mm^2 with 28 total veins meeting margin; and (7.5 mm long) 16.08 mm^2 total veins 37, respectively. *Hra nie* 3791/1147 (4.5 mm long) has area 5.6 mm^2 total veins 32. *Blattula mikro* 3791/5051 (4.8 mm long) area 5.7 mm^2 25 total veins. *Blattula universala* 3791/88 (9.0 mm long) area 17.40 mm^2 28 total veins; *Blattula mini* 516 (4.5 mm long) area 4.42 mm^2 28 total veins. *Raphidiomima chimnata* 3791/70 (17.7 mm long) area 58 mm^2 52 total veins, *Rhipidoblattina bakharensis* 3791/574 (14.0 mm long) area 61 mm^2 46 total veins; *Rhipidoblattina sisnerahkab* 3791/205 (14.8 mm long) area 34.67 mm^2; 48 total veins; *Rhipidoblattina sp.* 3791/117 (14.5 mm long) area 48.29 mm^2 45 total veins. These data are consistent with exponential relationships among the number of veins and forewing areas, with the reservation that this distribution is nearly linear ($P = 0.81$) in normal sized (not exceptionally small or exceptionally large) species (Table 4).

Taphonomy (see Table 5). 19 different pronota are represented at the bed. Additionally, pronota of *Policiblattula* and *Perlucipecta* are present, therefore at least 21 species were at this assemblage 275/1. This is the first known direct evidence

Table 4 Representation of formalised taxa, i.e., genera in major Jurassic sites with Cretaceous Chernovskie Kopi and Shar-Teg

	Bak	Kar	Mei	Zho	Hua	Xia	Dao	Dob	Min	Kho-Kho	Shu	Sha	Che
Ano	1	1	0	0	0	0	0	0	0	0	0	0	0
Blattula	1	1	1	1	1	1	1	1	1	1	1	1	1
Caloblattina	1	1	1	1	1	0	1	1	1	0	1	0	1
Dostavba	1	1	0	0	0	0	0	0	0	0	0	0	0
Hra	1	1	1	0	0	0	1	0	0	0	0	0	0
Nuurcala	1	1	1	0	0	0	1	0	0	0	0	0	0
Okras	1	0	0	0	0	0	0	0	0	0	0	0	0
Perlucipecta	1	1	1	1	0	0	1	0	0	0	1	0	0
Polliciblattula	1	1	0	0	0	0	0	0	0	0	0	0	0
Praeblattella	1	0	0	0	0	0	0	0	0	0	0	0	0
Raphidiomima	1	1	0	0	0	1	0	0	0	0	1	0	0
Rhipidoblattina	1	1	1	1	1	1	1	1	1	0	1	0	1
Solemnia	1	0	0	0	0	0	0	0	0	1	1	0	0
Truhla	1	0 ⍩	0	0	0	0	0	0	0	0	0	0	0
Chresmoda	1	1	0	0	0	0	1	0	0	0	0	0	0
Elisamoides	1	1	0	0	0	0	0	0	1	0	0	1	1
Euryblattula	0	1	0	0	0	1	0	1	0	0	0	0	0
Rhipidoblatta	0	1	1	0	1	0	1	1	1	0	0	0	1
Mongolblatta	0	1	0	0	0	0	0	0	0	0	0	1	1
Taublatta	0	0	0	0	0	0	0	0	0	0	1	0	0
Samaroblattula	0	0	0	1	0	0	0	0	0	0	1	0	0
Mesoblattula	0	0	0	0	0	0	0	0	0	0	1	0	0
Mesoblattina	0	1	0	1	0	0	0	1	0	0	0	0	0
Sogdoblatta	0	0	0	1	0	0	0	0	0	0	0	0	0
Irreblatta	0	0	0	0	0	0	0	0	0	1	0	0	0
Shartegoblattina	0	0	0	0	0	0	0	0	0	0	0	1	0
Elisama	0	1	0	0	0	0	0	0	0	0	0	1	0
Breviblattina	0	0	0	0	0	0	0	0	0	0	0	1	0
Kurablattina	0	0	0	0	0	0	0	0	1	0	0	0	0
Kurablattina	0	0	0	0	0	0	0	0	1	0	0	0	0
Divocina	0	0	0	0	0	0	1	0	0	0	0	0	0
Entropia	0	0	0	0	0	0	1	0	0	0	0	0	0
Fuzia	0	0	0	0	0	0	1	0	0	0	0	0	0
Parvifuzia	0	0	0	0	0	0	1	0	0	0	0	0	0
Colorifuzia	0	0	0	0	0	0	1	0	0	0	0	0	0

(continued)

Table 4 (continued)

	Bak	Kar	Mei	Zho	Hua	Xia	Dao	Dob	Min	Kho-Kho	Shu	Sha	Che
Fortiblatta	0	0	0	0	0	0	1	0	0	0	0	0	0
Artitocoblatta	0	1	0	0	0	0	0	0	0	0	0	0	0
Asioblatta	0	1	0	0	0	0	0	0	0	0	0	0	0
Decomposita	0	1	0	0	0	0	0	0	0	0	0	0	0
Karatavoblatta	0	1	0	0	0	0	1	0	0	0	0	0	0
Latiblatta	0	1	0	0	0	0	0	0	0	0	0	0	0
Paleovia	0	1	0	0	0	0	0	0	0	0	0	0	0
Rhipidoblattinopsis	0	1	0	0	0	0	0	0	0	0	0	0	0
Skok	0	1	0	0	0	0	0	0	0	0	0	0	0
Srdiecko	0	1	0	0	0	0	0	0	0	0	0	0	0
Falcatusiblatta	0	1	0	0	0	0	1	0	1	0	0	0	0

Genus *Elisamoides* is regarded for present in Bakhar, although specimens cannot be formalised

of the presence of more common species based on pronota (pronota are rarely preserved) rather than on preserved fore—and hindwings. The presence of numerous pronota at the assemblage might be interpreted as evidence for autochtonous preservation. Insects probably spent a considerable time in the water prior to sedimentation, enabling bacterians to dissolve articulations of body segments. Therefore, the partition of complete bodies is very low and consequently the separated body segments were not collected or represented in the collection. Taphonomy significantly influences diversity values as can be seen from the taphonomic differences paragraph above. Nevertheless, taphonomy barely affects the extremely high diversity of main beds, as this comparison is based on hundreds of Mesozoic localities, where only Karabastau has a comparable diversity. Thus, the diversity was actually extremely high.

Evidence above is presented for roughly coeval deposition of sediments in respective bed packages (208; 268; 275), and here is a presentation of **taphonomical differences within these packages**. 208/2 and 208/3 are taphonomically identical, species are identical and FW/HW partition comparable within this statistically insignificant set. Higher partition (4/12 compared with 1/56) of *Truhla vekov* is distinct, but also insignificant. 208/4 is taphonomically completely different, although represented with only four specimens, but those all represent other species as well as two immature individuals, which are all significant as only three immature individuals are contained in the whole Bakhar material.

275/1 and 275/2 also contain the same species. From those common species in much richer assemblage 275/1, *Caloblattina vremeni* is missing from 275/2. This absence is also not statistically significant (53/518 vs. 0/25), and this species additionally can be represented among nine unidentifiable forewings. A similar absence is *Blattula vulgara* (53/518 vs. 0/25), but Blattulidae are rare in the 275/2. It is too risky to attribute this difference to any taphonomical situation solely based on cockroaches.

268 contains assemblages 268/4, 8, 12, 14 and 19 are taphonomically totally consistent and exceptions are indigenous ($n = 1, 1$) *Dostavba pre* restricted to 268/19

Table 5 Taphonomy of 12 Bakhar assemblages

	203/6	208/2	208/3	208/4	268/4	268/8	268/12	268/14	268/19	275/1	275/2	328	SUM
Unidentificable pronota		1				1				38			40
Unidentificable FW			2					16 (2c)	2 (1c)	65	9		94 (3c)
Unidentificable HW			1				1	23	2	30			57
bodies										5	1		6
parts		8	3	1	4	6	1	23	4	44		2	96
larva				2				1					3
Ano da										87 (c2, fhp1, fh2, ffhh1, ffh1, 7hw)	3 (3f)		90 (c2, fhp1, fh2, ffhh1, ffh1, hw 7)
Ano net										2			2
Ano nym		10	1 (f)										11(1p, 1c,. 2 h)
Hra disko										3			3
Hra bavi					1					4			4
Hra nie						1 (hw)		3 (2hw)					5 (3hw)
Hra sp.								3					3 (3hw)
Blattula velika										14			14 (2c, 1H)

(continued)

Table 5 (continued)

	203/6	208/2	208/3	208/4	268/4	268/8	268/12	268/14	268/19	275/1	275/2	328	SUM
Blattula vulgara										53			53 (1 ffh, 1 bff, 1fh, 4 h, 1 bfovip)
Blattula mini										39 (1 h, 2c)	1		40(1 h, 2c)
Blattula micro	1									1			1
Blattula flamma		12											14
Blattula bacharensis												2	2
Blattula universala		3											3
Blattula anuniversala										2			2
Truhla vekov		1	4										5 (1 h)
Polliciblattula analis		1											1
Polliciblattula tatosanerata								1					1
Polliciblattula vana										1			1
Praeblattella jurassica												1	1
Perlucipecta cosmopoliana										4			4 (1c, 1 h)
Nuurcala cela				1?				1?		4			6

(continued)

Table 5 (continued)

	203/6	208/2	208/3	208/4	268/4	268/8	268/12	268/14	268/19	275/1	275/2	328	SUM
Rhipidoblattina bakharensis										21			21 (16f, 1c, 5hw)
Rhipidoblattina sisnerahkab					7 (1 h)	2 (1c)	15 (8f; 7hw)	195 (1 ffp, 1ffh, 1fh, 10c, 84f, 98 h)	22 (19fw, 13hw)				241 (11c, 1ffp, 1ffh, 1fh, 108f, 119 h)
Rhipidoblattina konserva					1 (c)	7 (1c, 1 h)	3 (2hw)	206 (96f; 3p; 80 h; 25c; 1fh; 1 complete)	14 (1c; 5 h)				231(1complete; 1fh; 141f; 88 h; 28c; 3p)
Okras sarko										2			2
Solemnia togokhudukhensis										33f (1ff, 1fh, 11c), 6hw, 1p	10(f)		50 (20f, 11c; 6hw; 1p; 1ff; 1fh)
Caloblattina vremeni										51			51 (33f, 11 h; 6c, 1p)
Dostavba pre									1				1
Raphidiomima chinnata		20											20 (14 f, 3c, 2 h, 1p).
Raphidiomima krajka										1 (c)			1 (c)
Raphidiomimidae sp.						1							1
Chresmoda sp.													

(continued)

Table 5 (continued)

	203/6	208/2	208/3	208/4	268/4	268/8	268/12	268/14	268/19	275/1	275/2	328	SUM
SUM	1	56	12	4	13	18	20	472	45	511	24	5	1182(1 (unassigned to a bed)

and one unformalised species represented with a single insignificant specimen (131 9/2.2 mm) restricted to 268/8. These doubtfully represent taphonomical exceptions and are likely stochastical in otherwise standard Bed 268/8 and Bed 268/19.

Deformities are present in five specimens (3791/606 hindwing R1-RS; 3791/88 forewing M-CuA; 3791/205 M-M incomplete fused vein), of which forewing specimen 706 posses double closely approximated fusions (CuA–CuA and adjacent CuA–M) and specimen 645 posses a tripple fusion on A1–CuP which is at the clavus margin and cannot be regarded as a deformity (Vršanský et al. 2017). It is among three complete specimens (12 wings), 57 completely drawn forewings (7 clavi) and 11 hindwings, which means documented partition of hard deformities (disregarding 645) is exactly 5%. This real number is even lower, because deformities were not recorded as distinct in the rest of the surveyed material. This parameter demonstrates stable evolution, while in diversification points this number reaches 20–100% (Vršanský et al. 2017). In the Middle to Late Jurassic context, this proportion is comparable only to Karabastau (1%), being significantly lower than in other sites including Daohugou (11.1%), further exemplifying extremely diverse and stable assemblages of Bakhar (and Karabastau).

General insect context is difficult to evaluate due to incomplete insect data. Pattern is distinct for uppermost strata 328, where in addition to *B. bacharensis*, a sister species to *B. choutinensis*, indigenous genera shared with Tithonian Khoutiin-Khotgor are present: *Haenbea badamgaravae* Popov, 1988 (Hemiptera: Corixidae) and "*Khoutynia*" (Odonata: Isophlebiidae). *Notocupes brachicephalus* Ponomarenko, 1994 also occurs in Cretaceous Shar-Teg. *Perlariopsis* Ping, 1928 (stonefly) (Ping 1928) occurs in Tithonian of Khoutiin-Khotgorin Mongolia but is also known from the Pliensbachian of Sagul in Kyrghyzstan. Other shared with Shar-Teg is the genus *Baga* Sukacheva, 1992 (finger-net caddisfly) (Sukacheva 1992). Genera with described taxa that are not indigenous include a beetle *Omma* Newman, 1839 known from Rhaetian of UK to Cenomanian of Myanmar (Carpenter 1992; Handlirsch 1906; Jarzembowski et al. 2016; Ponomarenko 1964, 1969, 1971; Soriano and Delclos 2006; Tan et al. 2006, 2012). *Tetraphalerus* Waterhouse, 1901 (see also Bouchard et al. 2011l; Carpenter 1992; Lin 1976, 1986; Lin 1992; Ponomarenko 1986, 2006; Ponomarenko et al. 2012; Soriano and Delclos 2006; Tan et al. 2007; Tan and Ren 2009 and Tan et al. 2012) occurs from Hettangian of Tonskiy, Kyrgyzstan to Eocene of Bembridge in the UK. A bug *Bakharia* Popov, 1988, also occurs in Upper Jurassic Kalgan Formation of Chita in Russia (Sukatsheva 1985); *Shuragobia* Popov, 1988, occurs in the Bajocian-Bathonian middle Jurassic of Mongolia (Popov 1988). *Olgamartynovia* Becker-Migdisova, 1958 (Becker-Migdisova 1958), spans from Hettangian of Kyrghystan and has LOD in Bakhar. *Baharellus* Storozhenko, 1988 ((Blattogryllidae), spans from the Triassic and has LOD in Bakhar (Storozhenko 1988). The orthopteran *Falsispeculum* Gorochov, 1985, occurs in Upper Jurassic of Karabastau formation (Gorochov 1985). *Mesotaeniopteryx* Martynov, 1937 (winter stonefly), (Martynov 1937) is known from the Pliensbachian of Shurab in Tajikistnan to Tithonian of *Kempendyay River in Russia*. *Xyelula* Rasnitsyn, 1969, a sawfly (Rasnitsyn 1969) is known from the Toarcian of Germany, England and India, to the Kimmeridgian of Karbastau in Kazakhstan. *Karabasia* Martynov, 1926 (true bug), occurs

(Martynov 1926) from the Shiti formation, Bajocian of Kazakhstan to the Ukurei formation, Hauterivian of Russia. *Permonka* cf. *sagulica* Rasnitsyn, 1977, relates to species from Pliensbachian of Shurab (Rasnitsyn 1977). *Cycloscytina* Martynov, 1926 (true bug), (Martynov 1926) from the Pliensbachian of Shurab has LOD in Bakhar. *Dysmorphoptila* Handlirsch, 1906 (true bug), (Handlirsch 1906) spans from the Rhaetian of UK to LOD in Bakhar. Only *Paranotonemoura* Cui et Béthoux in Cui et al. (2019; south forest flies) is shared with Daohugou. Not any cockroach is shared only with Daohugou, although *Ano, Blattula, Caloblattina, Hra, Liadoblattina, Nuurcala* and *Rhipidoblattina* occur also there.

Phylogenetically annotated character list (limited to forewings to avoid incompleteness of matrix and impossibility of Bayesian calculations; hindwings are known only for few new taxa; pronota bear little phylogenetic information) attributed to respective taxa in Table 2:

1. **Margins parallel**—*synapomorphy* of advanced cockroaches
2. **Wing base extended proximally**—*synapomorphy* of *H. bavi* and homoplasically others
3. **Shape significantly elongate**—*autapomorphy* of advanced Raphidiomimidae
4. **Shape not wide**—*synapomorphy* of advanced Mesozoic cockroaches, this *plesiomorphy* at the level of order was retained in certain Caloblattinidae
5. **Shape widened apically**—*synapomorphy* of certain Blattulidae
6. **Apex posed centrally**—*synapomorphy* of Corydioidea including Blattulidae
7. **Apex very sharp**—*autapomorphy* of *Raphidiomima chimnata*
8. **Apex round**—*synapomorphy* of certain Blattulidae and Liberiblattinidae
9. **Size very small**—*synapomorphy* of some Blattulidae
10. **Venation regular at margin**—*synapomorphy* of Phyloblattoidea and derived groups
11. **Venation regular also in R-M area**—*synapomorphy* of Corydioidea and modern groups
12. **Venation regular in clavus**—*synapomorphy* of advanced cockroaches including Corydioidea (also Blattulidae)
13. **Venation extremely reduced**—*autapomorphy* of *Polliciblattula*
14. **Venation moderately rich**—synapomorphy of advanced Mesozoic cockroaches; *plesiomorphically venation was very rich* at the level of order
15. **Main veins coloured**—*synapomorphy at unknown level*
16. **Main veins thick**—*synapomorphy* of certain Liberiblattinidae
17. **Intercalaries present**—*synapomorphy* of Phyloblattidae and derived groups; intercalaries are indistinct in advanced Mesoblattinidae
18. **Intercalaries thick**—*synapomorphy* of certain Liberiblattinidae
19. **Intercalaries coloured more pale compared to main veins**—*synapomorphy* of certain Liberiblattinidae and Blattulidae
20. **Intercalaries absent in clavus**—*synapomorphy* of advanced Mesozoic cockroaches, this *plesiomorphy* at the level of order retained in Caloblattinoidea
21. **Cross-veins present**—*synapomorphy* of Neorthroblattinidae and derived groups

22. **Membrane coloured**—*homoplasic* character along the taxonomic spectrum
23. **Coloration patterned (most simply as lines or stripes)**—*synapomorphy* of certain Caloblattinidae
24. **Coloration maculate or simply dotted**—*homoplasies* of *Okras* and *Raphidiomima*
25. **Coloration sophisticated (dots formed of stripes)**—*synapomorphy* of certain advanced Liberiblattinidae
26. **Coloration sporadical**—*autapomorphy*; homoplasic in numerous lineages
27. **Costal area narrow**—*synapomorphy* of advanced Liberiblattinidae
28. **Costal area shortened**—*synapomorphy* of some Blattulidae
29. **Costal area elongate**—*synapomorphy* of Liberiblattinidae
30. **SC simple**—*synapomorphy* of some Blattulidae, SC is plesiomorphically branched at the level of order
31. **SC sigmoidal**—*synapomorphy* of certain Corydioidea
32. **SC branched basally but not apically**—*synapomorphy* of advanced Mesozoic cockroaches, original state of terminal branches retained in some Caloblattinoidea but also in primitive Corydioidea
33. **R sigmoidally curved**—*synapomorphy* of Corydioidea
34. **R narrow**—*apomorphy*; R is plesiomorphically wide at the level of order
35. **R veins simple**—*synapomorphy* of advanced Mesozoic cockroaches; R is plesiomorphically tertiary branched
36. **RS not differentiated**—*apomorphy*, R is plesiomorphically differentiated at the level of order
37. **Basalmost R branched not branched extensively**—*synapomorphy* of advanced Mesozoic cockroaches, this plesiomorphy at the level of order retained in *Praeblattella*
38. **R not reaching apex**—*synapomorphy of* Liberiblattinidae
39. **M sharply descending from R stem**—*synapomorphy* of *Elisamoides* and some advanced Liberiblattinidae
40. **M and CuA not sigmoidal**—*synapomorphy* of Corydioidea
41. **M and CuA posteriorly curved apically**—*autapomorphy* of Raphidiomimidae
42. **Posteriormost CuA sigmoidal**—*synapomorphy* of certain Caloblattinidae and Raphidiomimidae
43. **Clavus small**—*autapomorphy* of advanced Raphidiomimidae; homoplasically present in some Blattulidae and Caloblattinidae
44. **Clavus with diagonal kink**—*synapomorphy* of certain Phyloblattidae and derived groups
45. **Clavus with dense cross-veins**—*synapomorphy* of certain Corydioidea and homoplasic in Raphidiomimidae
46. **Clavus sharply curved**—*synapomorphy* of certain Corydioidea
47. **Clavus extremely narrow**—*autapomorphy* of Raphidiomimidae
48. **Clavus anteriorly shortly cut**—*synapomorphy* of certain Corydioidea (including Blattulidae) and advanced living-type cockroaches (extremely expressed in Ectobiidae)

49. **Pseudovein present**—*synapomorphy* of certain Liberiblattinidae and mantodeans
50. **A simple**—*synapomorphy* of Blattulidae and homoplasically in modern groups
51. **A number reduced**—*synapomorphy of* Blattulidae
52. **A veins coloured**—*autapomorphy* of *R. krajka*
53. **A not sharply posteriorly curved**—*synapomorphy* of advanced cockroaches; Caloblattinidae and Raphidiomimidae retained the original state
54. **Distance between CuP and A1 insignificant**—*synapomorphy* of advanced Mesozoic cockroaches, this plesiomorphy at the level of Phyloblattidae was retained in Caloblattinoidea

Acknowledgements I thank Professor Alexandr Pavlovich Rasnitsyn (PIN RAS, Moscow) and Jun Hui Liang (Museum of Natural History, Tianjin) for their valuable reviews and advice. I deeply recognise the collective of the Artropod Laboratory of the Paleontological Institute of the Russian Academy of Sciences for collecting and revealing all specimens for study and for kind acceptance during the study. Especially acknowledged are Irina Dmitrievna Sukatcheva for extensive logistic support, Dmitry Evgenevich Shcherbakov, Dmitry Vladimirovič Vasilenko and Jan Hinkelman (IZ SAS Bratislava, Zagreb) for technical help and Alexandr Georgievich Ponomarenko for organising expeditions. I thank Tatiana Kúdelová (Comenius University, Bratislava) for advising phylogenetic analyses. This work was supported by the Slovak Research and Development Agency under the contracts no. APVV-0436-12; VEGA 2/0042/18 and by UNESCO-Amba/MVTS supporting grant of Presidium of the Slovak Academy of Sciences and its interacademic exchanges.

Correction to: Context of Bakhar Cockroaches

Correction to:
Chapter "Context of Bakhar Cockroaches" in: P. Vršanský,
Cockroaches from Jurassic sediments of the Bakhar
Formation in Mongolia, SpringerBriefs in Animal Sciences,
https://doi.org/10.1007/978-3-030-59407-7_5

In the original version of the book, Table 3 of Chap. 5 has been published with incorrect values. The table has been updated accordingly.

The updated version of this chapter can be found at
https://doi.org/10.1007/978-3-030-59407-7_5

Table 3 Phenetic similarity of major Jurassic sites with Cretaceous Chernovskie Kopi and Shar-Teg

	n	Bak	Kar	Mei	Zho	Hua	Xia	Dao	Dob	Min	Kho-Kho	Shu	Sha	Che
Shurab	37	6/9	6/9	4/9	5/9	3/9	3/9	6/9	4/9	3/9	2/9	–	1/9	2/9
Dobbertin	119	5/6	6/6	4/6	4/6	4/6	4/6	4/6	–	3/6	2/6	4/6	1/6	3/6
Mintaja	374	4/5	4/5	3/5	2/5	2/5	3/5	3/5	3/5	–	2/5	3/5	3/5	4/5
Daohugou	1500	8/16	3/16	5/16	3/16	4/16	3/16	–	4/16	3/16	2/16	6/16	1/16	3/16
Meitian	?	6/7	6/7	–	4/7	4/7	3/7	5/7	4/7	3/7	2/7	4/7	1/7	4/7
Zhouyingzi	?	4/7	5/7	4/7	–	3/7	2/7	3/7	4/7	2/7	2/7	5/7	2/7	2/7
Huapen	?	3/4	4/4	4/4	3/4	–	3/4	4/4	4/4	2/4	2/4	3/4	1/4	3/4
Xiaofanzhangzi	?	3/4	4/4	3/4	2/4	3/4	–	3/4	4/4	3/4	2/4	3/4	1/4	3/4
Bakhar	1179	–	11/16	6/16	4/16	3/16	3/16	8/16	5/16	4/16	2/16	6/16	2/16	4/16
Khoutiin-Khotgor	20	2/3	2/3	2/3	2/3	2/3	2/3	2/3	2/3	2/3	–	2/3	1/3	2/3
Karabastau	2175	11/77	–	6/77	5/77	4/77	4/77	3/77	6/77	4/77	2/77	6/77	4/77	4/77
Shar-Teg	170	2/6	4/6	1/6	2/6	1/6	1/6	1/6	1/6	3/6	1/6	1/6	–	6/6
Chernovskie Kopi	41	4/6	4/6	4/6	2/6	3/6	3/6	3/6	3/6	4/6	2/6	2/6	6/6	–

Monograph Zoobank:
9089B934-4F94-44E2-A188-1BDFC7EC2451

Zoobank codes for genera:

Ano Vršanský, 2020 06BF2E46-9357-4C1C-A82E-098CD61647EE
Blattula Handlirsch 71650847-ABE2-4BA7-B958-51B8A734321D
Caloblattina Handlirsch A371C30D-513F-41F0-9156-C3278C86950B
Dostavba Vršanský, 2020 879A432F-8894-43BD-BC1E-FC1555608048
Hra Vršanský, 2020 64A2F00D-562B-4419-95A4-9AE7162FF79F
Okras Vršanský, 2020 AACBE3B7-5C6E-4150-9AF7-B77DE8F9ABB2
Perlucipecta Wei et Ren 32127D86-934A-4EA7-BC04-22070BA7A3A4
Polliciblattula Vršanský, 2020 5FBF0C10-ECBC-4B40-998F-160F89B59595
Praeblattella Vršanský, 2020 86F555ED-7110-45DF-BF39-AEC66C77DC69
Raphidiomima Vishniakova 9BBEE43E-A852-4FB1-AB91-945998CA8323
Rhipidoblattina Handlirsch 17FAD62A-E799-4DD1-BDDE-9739BBEB0E31
Solemnia Vršanský, 2020 453B924E-1457-4027-AEEC-F1DFC49E5228
Truhla Vršanský, 2020 08ECEAC3-DE2C-470F-91A2-08CACE8D090B

Zoobank codes for species:

Ano da Vršanský, 2020 EF37F021-9315-4506-B871-4590449CC41D
Ano net Vršanský, 2020 3C71838C-12D2-4CAF-98E5-CDBD02A22F9D
Ano nym Vršanský, 2020 8BFE9028-AF00-4D8F-BE50-91867DC07BC8
Blattula anuniversala Vršanský, 2020 07928119-8792-420B-BE9A-A27B7AA59EF1
Blattula bacharensis Vršanský, 2020 FBAA94FB-A450-433D-A6FA-79951343B669
Blattula flamma Vršanský, 2020 76FF62A9-62A4-4B52-9C96-3B920B5ABB0E
Blattula micro Vršanský, 2020 EE220D89-FF6A-46AC-B722-F0D3887BC871
Blattula mini Vršanský, 2020 01BC5963-4E6C-4BD0-BD05-B23862957741
Blattula universala Vršanský, 2020 9F7322E5-4D9C-474D-9C26-AC5BD8D03CBB

P. Vršanský, *Cockroaches from Jurassic Sediments of the Bakhar Formation in Mongolia*, SpringerBriefs in Animal Sciences, https://doi.org/10.1007/978-3-030-59407-7

Blattula velika Vršanský, 2020 34EF1D69-DA71-44F8-B332-27A37028A075

Blattula vulgara Vršanský, 2020 F37CF2EE-0DC8-4C02-8C7C-82C1F30547D3

Caloblattina vremeni Vršanský, 2020 13947FB7-A080-416D-B9E0-560ECEBEA5D8

Dostavba pre Vršanský, 2020 CA13D2C6-FBAE-4595-BFC1-CE936A53F4F5

Hra bavi Vršanský, 2020 0376B1D0-89B2-405F-A8E5-5CDF9FB4B35F

Hra disko Vršanský, 2020 091B44EE-435E-43C3-959E-2A47FCF7D201

Hra nie Vršanský, 2020 015FB136-26F4-4672-B78E-192BB8CFFEB6

Nuurcala cela Vršanský, 2020 E8EE74E6-331F-4646-8438-D403E55491C3

Okras sarko Vršanský, 2020 414B99CE-A29D-43FC-A040-DD89DE09F52C

Perlucipecta cosmopolitana Vršanský, 2020 B69B101D-EA38-48E0-BF27-E8E3A6D3D6FE

Polliciblattula analis Vršanský, 2020 2730A8D0-FC8D-47F0-B9EE-85F773D64C22

Polliciblattula tatosanerata Vršanský, 2020 1768D4E0-9C61-4B5E-8F8D-746FB7A022E6

Polliciblattula vana Vršanský, 2020 137451D2-7CF4-49AC-BF81-505BB19034E2

Praeblattella jurassica Vršanský, 2020 F780734B-FCF2-43B2-9CA1-DC5EB6C35B4D

Raphidiomima chimnata Vršanský, 2020 B6FC59C5-9097-40E5-B4FC-E768CFDD2BB7

Raphidiomima krajka Vršanský, 2020 A9BE4A0B-F395-43C8-9EC8-1B270DC91249

Rhipidoblattina bakharensis Vršanský, 2020 7D95F62A-741E-4A19-BDA8-F32A9CE35A7C

Rhipidoblattina sisnerahkab Vršanský, 2020 60E91887-A842-4ACA-B6F3-B95EAE097969

Solemnia togokhudukhensis Vršanský, 2020 62B49809-C51A-4CE0-9618-286E8856928F

Truhla vekov Vršanský, 2020 68D1B826-55CA-4D62-97AC-76F3E90F831A

References

Anisyutkin LN, Gorochov AV (2008) A new genus and species of the cockroach family Blattulidae from Lebanese amber (Dictyoptera, Blattina). Paleontol J 42(1):43–46

Bai M, Beutel RG, Klass KD et al (2016) †Alienoptera—a new insect order in the roach-mantodean twilight zone. Gondwana Res 39:317–326. https://doi.org/10.1016/j.gr.2016.02.002

Bai M, Beutel RG, Zhang W et al (2018) A new Cretaceous insect with a unique cephalo-thoracic scissor device. Curr Biol 28:438–443. https://doi.org/10.1016/j.cub.2017.12.031

Bakhurina NN, Unwin DM (1995) A survey of pterosaurs from the Jurassic and Cretaceous of the former Soviet Union and Mongolia. Histor Biol 10:197–245

Barna P (2014) Low diversity cockroach assemblage from Chernovskie Kopi in Russia confirms wing deformities in insects at the Jurassic/Cretaceous boundary. Biologia 69:651–675. https://doi.org/10.2478/s11756-014-0349-9

Becker-Migdisova E (1958) Novye iskopaemye ravnokrylye. Materialy k Osnovam Paleontologii 2:57–67

Bode A (1953) Die Insektenfauna des Ostniedersachsischen Oberen Lias. Palaeontogr Abt A 103:1–375

Bouchard P, Bousquet Y, Davies AE et al (2011) Family-group names in Coleoptera (Insecta). ZooKeys 88:1–972. https://doi.org/10.3897/zookeys.88.807

Brauer F, Redtenbacher J, Ganglbauer L (1889) Fossile Insekten aus der Juraformation Ost-Sibiriens. Mem Acad Sci St Petersb, VII Série 36:1–22

Brunner von Wattenwyl K (1865) Nouveau Système des Blattaires. G. Braumüller, Vienne

Brunner von Wattenwyl K (1882) Prodromus der Europäischen Orthopteren. Monografien Entomologie, pp 1–466

Bryant D, Moulton V (2004) NeighborNet: an agglomerative algorithm for the construction of planar phylogenetic networks. Mol Biol Evol 21:255–265. https://doi.org/10.1007/3-540-45784-4_28

Carpenter FM (1992) Superclass hexapoda. In: Kaesler RL (ed) Treatise on invertebrate paleontology (Part R, Arthropoda 4, vol 3–4) The Geological Society of America, Boulder, pp 1–655

Cui Y, Ren D, Bethoux O (2019) The Pangean journey of 'south forestflies' (Insecta: Plecoptera) revealed by their first fossils. J Syst Palaeontol 17:255–268

Devyatkin EV, Martinson GG, Shuvalov VF et al (1975) Stratigraphy of Mesozoic deposits of western Mongolia. Trudy Sovmestnoj Rossiisko-Mongol'skoj Paleontologicheskoj Ekspeditsii 13:25–49

Deichmüller LV (1886) Die Insekten aus dem lithographischen Schiefer im Dresdner Museum. Mitt Konigl Miner-Geol-Praehist Museum 7:1–88

Engel MS, Pérez-de la Fuente R (2012) A new species of roach from the jurassic of India (Blattaria: Mesoblattinidae). J Kansas Entomol Soc 85:1–4. https://doi.org/10.2317/JKES110524.1

Fujiyama I (1973) Mesozoic insect fauna of East Asia. Part I. Introduction and Upper Triassic faunas. Bull Natn Sci Mus Tokyo 16:331–386

Fujiyama I (1974) A Liassic Cockroach from Toyora, Japan. Bull Natl Sci Mus 17:311–314

Fang Y, Zhang HC, Wang B et al (2013) A new cockroach (Blattaria: Caloblattinidae) from the Upper Triassic Xujiahe Formation of Sichuan Province, southwestern China. Insect Syst Evol 44:167–174. https://doi.org/10.1163/1876312X-44022101

Fu Q et al (2018) An unexpected noncarpellate epigynous flower from the Jurassic of China. Elife 7: AN: e38827. Doi 10.7554/eLife.38827

Gao T, Shih CK, Labandeira CC et al (2019) Maternal care by Early Cretaceous cockroaches. J Syst Palaeontol 17:379–391. https://doi.org/10.1080/14772019.2018.1426059

Germar EF (1839) Die versteinerten Insecten Solenhofens. Nova Acta Phys-Med Acad Caes Leop-Carol Nat Cur 19:187–222

Giebel CG (1856) Die Insecten und Spinnen der Vorwelt mit steter Berücksichtigung der lebenden Insekten und Spinnen. Die Fauna der Vorwelt 2:1–511

Geinitz FE (1880) Der Jura von Dobbertin in Mecklenburg und seine Versteinerungen. Z Dtsch Geol Ges 32:510–535

Geinitz FE (1883) Die Flötzformationen Mecklenburgs. Arch Ver Fr Naturges Mecklenburg 37:1–151

Geinitz FE (1884) Ueber die Fauna des Dobbertiner Lias. Z Dtsch Geol Ges 36:566–583

Geinitz FE (1887) Neue Aufschlusse der Flozformation Mecklenburgs. IX Beitrag zur Geologie Mecklenburgs. IV Jura. Arch Ver Fr Naturges Mecklenburg 41:143–216

Gorochov AV (1985) Mesozoic crickets (Orthoptera, Grylloidea) of Asia. Paleontol J 19:56–66

Guo YX, Ren D (2011) A new cockroach genus of the family Fuziidae from northeastern China (Insecta: Blattida). Acta Geol Sin 85:501–506

Handlirsch A (1906 -1908) Die Fossilen Insekten und die Phylogenie der Rezenten Formen. Ein Handbuch fur Palaontologen und Zoologen. Verlag von Wilhelm Engelmann, Leipzig

Handlirsch A (1939) Neue Untersuchungen über die fossilen Insekten mit Ergänzungen und Nachträgen sowie Ausblicken auf phylogenetische, palaeogeographische und allgemein biologische Probleme. II Teil. Ann Nat Hist Mus Wien 49:1–240

Heer O (1852) Die Lias-Insel des Aargau's. Zwei Geologische Vorträge Gehalten im März 1852 1:1–15

Heer O (1864) Ueber die fossilen Kakerlaken. Vierteljahresschr Naturforsch Ges Zürich 9:273–302

Heer O (1865) Die Urwelt der Schweiz. Friedrich Schulthess, Zurich

Haughton SH (1924) The fauna and stratigraphy of the Stormberg Series. Ann S Afr Mus 12:323–497

Hinkelman J (2019) *Spinaeblattina myanmarensis* gen. et sp. nov. and *Blattoothecichnus argenteus* ichnogen. et ichnosp. nov. (both Mesoblattinidae) from mid-Cretaceous Myanmar amber. Cretac Res 99:229–239. https://doi.org/10.1016/j.cretres.2019.02.026

Hinkelman J (2020) Earliest behavioral mimicry and possible food begging in a Mesozoic alienopterid pollinator. Biologia 75:83–92. https://doi.org/10.2478/s11756-019-00278-z

Hinkelman J, Vršanská L (2020) A Myanmar amber cockroach with protruding feces contains pollen and a rich microcenosis. Sci Nat 107:13. https://doi.org/10.1007/s00114-020-1669-y

Huber I (1974) Taxonomic and ontogenetic studies of cockroaches (Blattaria). Univ Kans Sci Bull 50:233–332

Hong YC (1980) New genus and species of Mesoblattinidae (Blattodea, Insecta) in China. Bul Sin Acad Sci Geol Sci VI 1:49–60

Hong YC (1982) Mesozoic fossil insects of Jiuquan Basin in Gansu Province. Geological Publishing House, Beijing

Hong YC (1983) Middle Jurassic fossil insects in North China. Geological Publishing House, Beijing

Hong YC (1986) New fossil insects of Haifanggou formation, Liaoning Province. J Changchun Col Geol 4:10–16

Hong YC (1987) The study of Early Cretaceous insects of Kezuo, west Liaoning. Prof Papers Stratigr Palaeontol 18:76–87

Hong YC (1997) Fossil Blattaria from the Houcheng formation at Yanqing County Beijing. Beijing Geology 9:1–6

Hong YC, Xiao ZZ (1997) New fossil Blattodea, Coleoptera and Mecoptera (Insecta) from Houcheng Formation of Yanqing County, Beijing. Beijing Geol 1997:1–10

Huber P, McDonald NG, Olsen PE (2003) Early Jurassic insects from the Newark Supergroup, Northeastern United States. In: LeTourneau PM, Olsen PE (eds) The Great Rift Valleys of Pangea in Eastern North America. Sedimentology, stratigraphy, and paleontology. Columbia University Press, New York, 2: 206–223

Chen T, Liu SH, Le XJ, Chen L (2019) A new cockroach (Insecta: Blattaria: Blattulidae) from the Lower Cretaceous Laiyang Formation of China. Cretac Res 101:17–22

Jarzembowski EA, Wang B, Zheng DR (2016) An amber double first: a new brochocolein beetle (Coleoptera: Archostemata) from northern Myanmar. Proc Geol Assoc 127:676–680

Katinas V (1983) Baltijos gintaras. Mokslas, Vilnius

Kostina EI, Herman AB, Kodrul TM (2015) Early Middle Jurassic (possibly Aalenian) Tsagan-Ovoo Flora of Central Mongolia. Rev Palaeobot Palyno 220:44–68. https://doi.org/10.1016/j.rev palbo.2015.04.010

Kostina EI, Herman AB (2016) Middle Jurassic floras of Mongolia: Composition, age, and phytogeographic position. Paleontol J 50:1437–1450. https://doi.org/10.1134/S00310301161 20108

Kaddumi HF (2005) Amber of Jordan: the oldest prehistoric insects in fossilized resin. Contributions from the Eternal river Museum of Natural History, Amman

Kočárek P (2018) Alienopterella stigmatica gen. et sp. nov.: the second known species and specimen of Alienoptera extends knowledge about this Cretaceous order (Insecta: Polyneoptera). J Syst Palaeontol 17:491–499. https://doi.org/10.1080/14772019.2018.1440440

Kočárek P (2018) The cephalothoracic apparatus of Caputoraptor elegans may have been used to squeeze prey. Curr Biol 28:803–825. https://doi.org/10.1016/j.cub.2018.06.046

Latreille PA (1810) Considerations generales sur l'ordre naturel des animaux composant les classes des crustaces, des arachnides, et des insectes, avec un tableau methodique de leurs genres, disposes en familles. Chez. F, Schoell, Paris

Lee SW (2016) Taxonomic diversity of cockroach assemblages (Blattaria, Insecta) of the Aptian Crato Formation (Cretaceous, NE Brazil). Geol Carpath 67:433–450

Li XR, Huang D (2018) A new cretaceous cockroach with heterogeneous tarsi preserved in Burmese amber (Dictyoptera, Blattodea, Corydiidae). Cretac Res 92:12–17. https://doi.org/10.1016/j.cre tres.2018.07.017

Li X, Huang D (2019) A new mid-Cretaceous cockroach of stem Nocticolidae and reestimating the age of Corydioidea (Dictyoptera: Blattodea). Cretac Res 106:104202. https://doi.org/10.1016/j. cretres.2019.104202

Lin QB (1978) On the fossil Blattoidea of China. Acta Entomol Sin 21:335–342

Lin QB (1982) Paleontological Atlas of northwest China Shaanxi Gansu Ningxia volume Part III Mesozoic and Cenozoic. Geological Publishing House, Beijing

Lin QB (1985) Insect Fossils from the Hanshan Formation at Hanshan County, Anhui Province. Acta Palaeontol Sin 24:300–315

Lin QB (1986) Early Mesozoic fossil insects from South China. Palaeontol Sin B 170:1–112

Lin QB (1992) Late Triassic insect fauna from Toksun, Xinjiang. Acta Palaeontol Sin 31:313–335

Liang J, Vršanský P, Ren D et al (2009) A new Jurassic carnivorous cockroach (Insecta, Blattaria, Raphidiomimidae) from the Inner Mongolia in China. Zootaxa 1974:17–30

Liang JH, Vršanský P, Ren D (2012) Variability and symmetry of a Jurassic nocturnal predatory cockroach (Blattida: Raphidiomimidae). Rev Mex Cienc Geol 29:411–421

Liang J, Shih C, Wang L et al (2018) New Jurassic predatory cockroaches (Blattaria: Raphidiomimidae) from Daohugou, China and Karatau Kazakhstan. Alcheringa 42(1):101–109

Liang J, Shih C, Wang L et al (2019) New cockroaches (Insecta, Blattaria, Fuziidae) from the Middle Jurassic Jiulongshan formation in northeastern China. Alcheringa 43:441–448. https://doi.org/10.1080/03115518.2019.1576061

Martin SK (2010) Early Jurassic cockroaches (Blattodea) from the Mintaja insect locality, Western Australia. Alavesia 3:55–72

Martínez-Delclós X (1993) Blátidos (Insecta, Blattodea) del Cretácico Inferior de España. Familias Mesoblattinidae, Blattulidae y Poliphagidae. Boletín Geológico y Minero 104–538

Martynov AV (1926) Jurassic fossil Insect from Turkestan. 6. Homoptera and Psocoptera. Izv Akad Nauk 20:1349–1366

Martynov AV (1937) Liassic insects from Shurab and Kisyl-Kiya, Part I, Various orders except Blattodea and Coleoptera. Trudy Paleontol Inst Akad Nauk SSSR 7:1–178

Martynova OM (1951) Dva novykh nizhneleyasovykh vida nasekomykh iz kizil-kii (Kirgizskaya SSR). Dokl Akad Nauk SSSR78:1009–1011

Mlynský T, Wu H, Koubová I (2019) Dominant Burmite cockroach *Jantaropterix ellenbergeri* sp.n. might laid isolated eggs together. Palaeontogr Abt A 314:69–79. https://doi.org/10.1127/pala/2019/0091

Oppenheim P (1888) Die Insectenwelt des lithographischen Scheifers in Bayern. Palaeontogr 35:215–254

Oružinský R, Vršanský P (2017) Cockroach forewing area and venation variabilities relate. Biologia 72(7):814–818

Papier F, Grauvogel-Stamm L, Nel A (1994) *Subioblatta undulata* n. sp., une nouvelle blatte (Subioblattidae Schneider) du Buntsandstein supérieur (Anisien) des Vosges (France). Morphologie, systématique et affinités. N Jb für Geol Paläont Mh 1994:277–290

Papier F, Grauvogel-Stamm L (1995) Les Blattodea du Trias: Le genre *Voltziablatta* n. gen. du Buntsandstein supérieur des Vosges (France). Palaeontogr Abt A 235:141–162

Ping C (1928) Study of the Cretaceous fossil insects of China. Palaeontol Sin B 13:5–57

Podstrelená L, Sendi H (2018) *Cratovitisma* Bechly, 2007 (Blattaria: Umenocolidae) recorded in Lebanese and Myanmar ambers. Palaeontogr Abt A 310:121–129. https://doi.org/10.1127/pala/2018/0076

Poinar GO (1992) Life in amber. Stanford University Press, Palo Alto

Poinar GO, Buckley R (2006) Nematode (Nematoda: Mermithidae) and hairworm (Nematomorpha: Chordodidae) parasites in early Cretaceous amber. J Invertebr Pathol 93:36–41. https://doi.org/10.1016/j.jip.2006.04.006

Poinar GO (2009) *Meloe dominicanus* n. sp. (Coleoptera: Meloididae) phoretic on the bee Proplebia dominicana (Hymenoptera: Apidae) in Dominican amber. Proc Entomol Soc Wash 111:145–150

Poinar GO (2009) Description of an Early Cretaceous termite (Isoptera: Kalotermitidae) and its associated intestinal protozoa, with comments on their coevolution. Parasite Vector 2:12. https://doi.org/10.1186/1756-3305-2-12

Poinar GO, Brown AE (2017) An exotic insect *Aethiocarenus burmanicus* gen. et sp. nov. (Aethiocarenodea ord. nov., Aethiocarenidae fam. nov.) from mid-Cretaceous Myanmar amber. Cretac Res 72:100–104. https://doi.org/10.1016/j.cretres.2016.12.011

Ponomarenko AG (1964) New beetles of the family Cupedidae from the Jurassic of Karatau. Paleontol J 2:49–61

Ponomarenko AG (1969) Historical development of Archostematan beetles. Trudy Paleontol Inst Akad Nauk SSSR 125:1–240

Ponomarenko AG (1971) On the taxonomic position of some beetles from the Solenhofen shales of Bavaria. Paleontol J 5:62–75

Ponomarenko AG (1986a) Insects in the Early Cretaceous ecosystems of West Mongolia. In: Rasnitsyn AP (ed) Nasekomye rannemelovykh ekosistemakh Zapadonoy Mongolii. Sovmestnaya Rossiisko-Mongol'skaya Paleontologicheskaya Ekspeditsiya, Nauka, Moscow, 28: 183–201

Ponomarenko AG (1986b) Scarabaeiformes *incertae sedis*. In: Rasnitsyn AP (ed) Nasekomye rannemelovykh ekosistemakh Zapadonoy Mongolii. Sovmestnaya Rossiisko-Mongol'skaya Paleontologicheskaya Ekspeditsiya. Nauka, Moscow, 28: 110–112

Ponomarenko AG (2006) On the types of Mesozoic archostematan beetles (Insecta, Coleoptera, Archostemata) in the Natural History Museum, London. Paleont J 40:90–99. https://doi.org/10.1134/S0031030106010102

Ponomarenko AG (2019) Bakhar. http://palaeoentomolog.ru/Collections/loc_b.html#bahar. Accessed 2019-11-17

Ponomarenko AG, Popov YA (2016) Paleoentomology of Mongolia. Paleontol J 50:1390–1400

Ponomarenko AG, Yan EV, Wang B et al (2012) Revision of some early Mesozoic beetles from China. Acta Palaeontol Sin 51:475–490

Popov A (1988) Novye Mezozojskie Klopy Grebljaki (Corixidae, Shurabellidae). Novye Iskopaemykh Bespozvonochnykh Mongolii, Sovmestnaya Sovetsko-Mongol'skaya Paleontologicheskaya Ekspeditsiya 33:63–71

Qiu L, Wang ZQ, Che YL (2019a) A new corydiid cockroach with large holoptic eyes in Upper Cretaceous Burmese amber (Blattodea: Corydiidae: Euthyrrhaphinae). Cretac Res 96:179–183. http://dx.doi.org/10.1016/j.cretres.2018.12.018

Qiu L, Wang ZQ, Che YL (2019) First record of Blattulidae from mid-Cretaceous Burmese amber (Insecta: Dictyoptera). Cretac Res 99:281–290. https://doi.org/10.1016/j.cretres.2019.03.011

Qiu L, Liu YC, Wang ZQ et al (2020) The first blattid cockroach (Dictyoptera: Blattodea) in Cretaceous amber and the reconsideration of purported Blattidae. Cretac Res 109:104359. https://doi.org/10.1016/j.cretres.2019.104359

Rasnitsyn AP (1969) Proiskhozhdenie i evolyutsiya nizshikh pereponchatokrylykh. Trudy Paleontol Inst Akad Nauk SSSR 123:1–196

Rasnitsyn AP (1977) New Paleozoic and Mesozoic insects. Paleontol J 11:60–72

Ren D, Lu LW, Ji SA et al (1995) Fauna and stratigraphy of Jurassic-Cretaceous in Beijing and adjacent areas. Seismic Publishing House, Beijing

Ren D (1995) Systematic Palaeontology. Fossil Insects. In: Ren D, Lu LW, Guo ZG, Ji SA (eds) Faunae and stratigraphy of Jurassic-Cretaceous in Beijing and the adjacent areas. Seismic Publishing House, Beijing, pp 47–121

Ross A, Grimaldi D (2004) *Raphidiomimula*, an enigmatic new cockroach in Cretaceous amber from Myanmar (Burma) (Insecta: Blattodea: Raphidiomimidae). J Syst Palaeontol 2:101–104. https://doi.org/10.1017/S1477201904001142

Saussure H (1864) Memoires de la Société de physique et d'histoire naturelle de Genève. Société de physique et d'histoire naturelle de Genève 18:230

Sendi H, Azar D (2017) New aposematic and presumably repellent bark cockroach from Lebanese amber. Cretac Res 72:13–17. http://dx.doi.org/10.1016%2Fj.cretres.2016.11.013

Sendi H, Vršanský P, Podstrelená L, Hinkelman J, Kúdelová T, Kúdela M, Vidlička Ľ, Ren XY, Quicke DLJ (2020) Nocticolid cockroaches are the only known dinosaur age cave survivors. Gondwana Research 82:288–298

Sendi H, Hinkelman J, Vrsanska L, Kudelova T, Kudela M, Zuber M, van de Kamp T, Vrsansky P (2020) Roach nectarivory, gymnosperm and earliest flower pollination evidence from Cretaceous ambers. Biologia. https://doi.org/10.2478/s11756-019-00412-x

Scudder SH (1886) A review of Mesozoic cockroaches. Memoirs Boston Soc Nat'l Hist 3:439–484

Schneider J (1977) Zur Variabilität der Flügel paläozoischer Blattodea (Insecta), Teil I. Freiberg Forsch H C 326:87–105

Schneider J (1978) Zur Variabilität der Flügel paläozoischer Blattodea (Insecta), Teil II. Freiberg Forsch H C 334:21–39

Schneider J (1978) Zur Taxonomie und Biostratigraphie der Blattodea (Insecta) des Karbon und Perm der DDR. Freiberg Forsch H C 340:1–152

Schneider J (1978) Revision der Poroblattinidae (Insecta, Blattodea) des europäischen und nordamerikanischen Oberkarbon und Perm. Freiberg Forsch H C 342:55–66

Schneider J (1980) Zur Entomofauna des Jungpaläozoikums der Boskovicer Furche (CSSR), Teil 1: Mylacridae (Insecta, Blattodea). Freiberg Forsch H C 357:43–55

Schneider J (1982) Entwurf einer Zonengliederung für das euramerische Permokarbon mittels der Spiloblattinidae (Blattodea, Insecta). Freiberg Forsch H C 375:27–47

Schneider J (1983) Die Blattodea (Insecta) des Paläozoikums, Teil 1: Systematik, Ökologie und Biostratigraphie. Freiberg Forsch H C 382:106–145

Schneider J (1984) Die Blattodea (Insecta) des Paläozoikums, Teil 2: Morphogenese der Flügel-strukturen und Phylogenie. Freiberg Forsch H C 391:5–34

Schneider J (1984) Zur Entomofauna des Jungpaläozoikums der Boskovicer Furche (CSSR), Teil 2: Phyloblattidae (Insecta, Blattodea). Freiberg Forsch H C 395:19–37

Schneider J, Werneburg R (1993) Neue Spiloblattinidae (Insecta, Blattodea) aus dem Oberkarbon und Unterperm von Mitteleuropa sowie die Biostratigraphie des Rotliegend. Veröffentlichungen Naturhistorisches Museum Schleusingen 7(8):31–52

Sinitsa SM (1993) Jurassic continental biocenoses of southern Siberia and adjacent territories Trudy Paleontol. Inst Akad Nauk SSSR 213:25–36

Soriano C, Delclòs X (2006) New cupedids from the lower Cretaceous of Spain and the palaeogeography of the family. Acta Palaeontol Pol 51:185–200

Storozhenko Y (1988) New and little-known Mesozoic grylloblattids (Insecta). Paleontol J 22:45–52

Sukatsheva ID (1985) Yurskie rucheyniki yuzhnoy Sibiri. In: Rasnitsyn AP (ed) Yurskie Nasekomye Sibiri i Mongolii. Trudy Paleontol Inst Akad Nauk SSSR, Moskva, pp 115–119

Sukatsheva ID (1992) Novye iskopaemye predstaviteli otryada rucheynikov (Phryganeida) iz Mongolii. Novye Taksony Iskopaemykh Bespozvonochnykh Mongolii, Sovmestnaya Rossiisko-Mongol'skaya Paleontologicheskaya Ekspeditsiya 4:111–117

Sukatsheva ID (1994) Upper Jurassic caddis cases (Trichoptera) from Mongolia. Paleontol J 4:76–85

Swofford DL (2003) PAUP*. Phylogenetic analysis using parsimony (*and Other Methods). Ver. 4. Sinauer Associates, Sunderland

Šmídová L, Lei X (2017) The earliest amber-recorded type cockroach family was aposematic (Blattaria: Blattidae). Cretac Res 72:189–199. https://doi.org/10.1016/j.cretres.2017.01.008

Šmídová L (2019) Unusual cockroaches (Blattidae) from Cenomanian Myanmar amber. Dissertation, Univerzita Karlova v Praze

Šmídová L (2020) Cryptic bark cockroach (Blattinae: Bubosa poinari gen. et sp.nov.) from mid-Cretaceous amber of northern Myanmar. Cretac Res 109:104383 https://doi.org/10.1016/j.cretres.2020.104383

Tan JJ, Ren D, Shih CK et al (2006) New fossil beetles of the family Ommatidae from Jehol Biota of China (Coleoptera: Archostemata). Acta Geol Sin 80:474–485

Tan JJ, Ren D, Shin C (2007) New beetles (Insecta, Coleoptera, Archostemata) from the late Mesozoic of North China. Ann Zool 57:231–247. https://doi.org/10.1016/j.crpv.2013.06.001

Tan JJ, Ren D (2009) Mesozoic Archostematan Fauna from China. Science Press, Beijing

Tan JJ, Wang Y, Ren D et al (2012) New fossil species of ommatids (Coleoptera: Archostemata) from the Middle Mesozoic of China illuminating the phylogeny of Ommatidae. BMC Evol Biol 12:113. https://doi.org/10.1186/1471-2148-12-113

Vidlička Ľ (2001) Fauna Slovenska Blattaria—šváby Mantodea—Modlivky (Insecta: Orthopteroidea). Veda vydavateľstvo SAV, Bratislava

Vishnyakova VN (1968) Mezozoiskiye Tarakany s Naruzhnym Yaytsekladom i Osobennosti ikh Rasmnozheniya (Blattodea) [Mesozoic cockroach ovipositors and the peculiarity of their reproduction (Blattodea)]. Yurskye Nasekomiye Karatau [Jurassic Insects of Karatau] 55–86. Nauka, Moscow, p 252

Vishnyakova VN (1971) Structure of the abdominal appendages of the Mesozoic roaches (Insecta: Blattodea). In: Obruchev DV, Shimansky VN (eds) Current problems in palaeontology. Trudy Paleontol Inst Akad Nauk SSSR 130. Nauka, Moscow, p 174–186

Vishnyakova VN (1973) New roaches (Insecta: Blattodea) from the Upper Jurassic deposits of Karatau Range. In: Narchuk EP (ed) Problems of the Insect Palaeontology Lectures on the XXIV Annual Readings in Memory of N. A. Kholodkovsky (1–2 April, 1971). Nauka, Leningrad, pp 64–77

Vishnyakova VN (1982) Jurassic cockroaches of the new family Blattulidae from Siberia. Paleontol J 16:67–77

Vishnyakova VN (1983) Jurassic cockroaches of the Mesoblattinidae family from Siberia. Paleontol J 17:76–90

Vishnyakova VN (1985) Tarakany (Blattida=Blattodea) Yuri yuzhnoy Sibiri i zapadnoy Mongolii. Yurskie Nasekomye Sibiri i Mongolii 138–147

Vishnyakova VN (1986) Blattida (=Blattodea). In: Rasnitsyn (ed) Insects in the Early cretaceous ecosystems of West Mongolia. Trans. Joint Soviet-Mongolian Pal. Exp. Moscow nauka 28:166–169

Vishnyakova VN (1993) New Paleozoic Spiloblattinidae from Russia. Paleontol J 27:135–147

Vakhrameev VA (1991) Jurassic and Cretaceous floras and climates of the Earth. Cambridge University Press, Cambridge

Vršanský P (2000) Decreasing variability-from the Carboniferous to the present! (validated on independent lineages of Blattaria). Paleontol J 34:374–379

Vršanský P (2002) Origin and the early evolution of mantises. AMBA Projekty 6:1–16

Vršanský P (2003) Umenocoleoidea—an amazing lineage of aberrant insects (Insecta, Blattaria). AMBA Projekty 7:1–32

Vršanský P (2003) Unique assemblage of Dictyoptera (Insecta-Blattaria, Mantodea, Isoptera, Mantodea) from the Lower Cretaceous of Bon Tsagaan Nuur in Mongolia. Entomological Problems 33:119–151

Vršanský P (2004) Transitional Jurassic/Cretaceous Cockroach assemblage (Insecta, Blattaria) from the Shar-Teg in Mongolia. Geol Carpath 55:457–468

Vršanský P (2005) Mass mutations of insects at the Jurassic/Cretaceous boundary. Geol Carpath 56:473–781

Vršanský P, Ansorge J (2007) Lower Jurassic cockroaches (Insecta: Blattaria) from Germany and England. Afr Invertebr 48:103–126

Vršanský P (2007) Jumping cockroaches (Blattaria, Skokidae fam.n.) from the late Jurassic of Karatau in Kazakhstan. Biologia 62:588–592

Vršanský P, Liang JH, Ren D (2009) Advanced morphology and behaviour of extinct earwig-like cockroaches (Blattida: Fuziidae fam. nov.). Geol Carpath 60:449–462

Vršanský P (2009) Albian cockroaches (Insecta, Blattida) from French amber of Archingeay. Geodiversitas 31:73–98. https://doi.org/10.5252/g2009n1a7

Vršanský P (2010) Cockroach as the earliest eusocial animal. Acta Geol Sin 84:793–808. https://doi.org/10.1111/j.1755-6724.2010.00261.x

Vršanský P, Liang J, Ren D (2012) Malformed cockroach (Blattida:Liberiblattinidae) in the Middle Jurassic sediments from China. Orient Insects 46:12–18. https://doi.org/10.1080/00305316.2012.675482

Vršanský P, Vidlička L, Barna P et al (2013a) Paleocene origin of the cockroach families Blaberidae and Corydiidae: Evidence from Amur River region of Russia. Zootaxa 3635:117. https://doi.org/10.11646/zootaxa.3635.2.2.

Vršanský P, van de Kamp T, Azar D et al (2013) Cockroaches probably cleaned up after dinosaurs. PLoS One 8:e80560. https://doi.org/10.1371/journal.pone.0080560

Vršanský P, Bechly G (2015) New predatory cockroaches (Insecta: Blattaria: Manipulatoridae fam. n.) from the Upper Cretaceous myanmar amber. Geol Carpath 66:133–138. https://doi.org/10.1515/geoca-2015-0015

Vršanský P, Wang B (2017) A new cockroach, with bipectinate antennae (Blattaria: Olidae fam. nov.) further highlights the differences between the burmite and other faunas. Biologia 72:1327–1333. https://doi.org/10.1515/biolog-2017-0144

Vršanský P, Bechly G, Zhang Q et al (2018) Batesian insect-insect mimicry-related explosive radiation of ancient alienopterid cockroaches. Biologia 73:987–1006. https://doi.org/10.2478/s11756-018-0117-3

Vršanský P, Oružinský R, Aristov D et al (2017) Temporary deleterious mass mutations relate to originations of cockroach families. Biologia 72:886–912. https://doi.org/10.1515/biolog-2017-0096

Vršanský P, Šmídová L, Sendi H et al (2019) Parasitic cockroaches indicate complex states of earliest proved ants. Biologia 74:65–89. https://doi.org/10.2478/s11756-018-0146-y

Vršanský P, Sendi H, Aristov D et al (2019) Ancient roaches further exemplify 'no land return' in aquatic insects. Gondwana Res 68:22–33. https://doi.org/10.1016/j.gr.2018.10.020

Vršanský P, Vishnyakova VN, Rasnitsyn AP (2002) Order Blattida Latreille, 1810. In: Rasnitsyn AP (ed) History of insects. Springer, Netherlands, pp 263–270

Vršanský P, Vršanská L, Beňo M et al (2019) Pathogenic DWV infection symptoms in a Cretaceous cockroach. Palaeontogr Abt A 314:1–10. https://doi.org/10.1127/pala/2019/0084

Vršanský P, Koubová I, Vršanská L et al (2019) Early wood-boring 'mole roach' reveals eusociality "missing ring". AMBA projekty 9:1–28

Whalley PES (1985) The systematics and palaeogeography of the Lower Jurassic insects of Dorset, England. Bull Br Mus Nat Hist Geol 39:107–189

Wang WL (1987) Early Mesozoic insects fossils from western Liaoning. Mesozoic Stratigr Palaeontol West Liaon 202–222

Wang TT, Liang JH, Ren D (2007) Variability of *Habroblattula drepanoides* gen. et. sp nov (Insecta : Blattaria: Blattulidae) from the Yixian Formation in Liaoning, China. Zootaxa 1443:17–27

Wei DD, Shih CK, Ren D (2012) *Arcofuzia cana* gen. et sp. n. (Insecta, Blattaria, Fuziidae) from the Middle Jurassic sediments of Inner Mongolia. China. Zootaxa 3597:25–32

Wei DD, Liang JH, Ren D (2012) A new species of Fuziidae (Insecta, Blattida) from the Inner Mongolia, China. ZooKeys 217:53–61

Wei DD, Liang JH, Ren D (2013) A new fossil genus of Fuziidae (Insecta, Blattida) from the Middle Jurassic of Jiulongshan Formation, China. Geodiversitas 35:335–343. https://doi.org/10.5252/g2013n2a3

Zherikhin VV (1985) Jurassic continental biocenoses of southern Siberia and adjacent territories. Trudy Paleontol Inst Akad Nauk SSSR 213:100–131

Zhang JF (1986) Some fossil insects from the Jurassic of northern Hebei, China. Paleontol Stratigr Shandong 74–84

Printed in the United States
By Bookmasters